THE
RATIONAL FRONTIERS
OF SCIENCE

THE RATIONAL FRONTIERS OF SCIENCE

The Origins of Knowledge and Explanation in Science

M. Rocha e Silva

KRIEGER PUBLISHING COMPANY
MALABAR, FLORIDA
1982

First Edition 1982
Based upon Title of the Brazilian Edition:
Ciência Pura, Ciência Aplicada (1976)

Printed and Published by
ROBERT E. KRIEGER PUBLISHING COMPANY, INC.
KRIEGER DRIVE
MALABAR, FL 32950

Copyright © 1982 by
Robert E. Krieger Publishing Company, Inc.

All rights reserved. No part of this book may be reproduced in any form or by any electronic or mechanical means including information storage and retrieval systems without permission in writing from the publisher.

Printed in the United States of America

Library of Congress Cataloging in Publication Data
Silva, Maurício Rocha e, 1910-
   The rational frontiers of science.

   Translation of: Ciência pura, ciênsa aplicada.
   Bibliography: p.
   Includes index.
   1. Science.   I. Title.
Q172.S5413   500   82-46
ISBN 0-89874-190-4   AACR2

Contents

INTRODUCTION: Pure Science and Applied Science — 1

CHAPTER I: Science and the Man — 13

CHAPTER II: Dialectics and the Scientific Method — 29

CHAPTER III: Origins of Scientific Knowledge — 45

CHAPTER IV: The Origins of Scientific Explanation — 65

CHAPTER V: The Language of Living Matter — 91

References and Additional Bibliography (AB) — 111

Index of Names — 117

Introduction

PURE SCIENCE AND APPLIED SCIENCE

Introduction

PURE SCIENCE AND APPLIED SCIENCE

We live in an age in which the value of science is being questioned by philosophers, politicians and laymen in spite of the wonder with which its practical and theoretical achievements are regarded. Scientists themselves show ambivalence about the real importance of scientific investigation, notwithstanding the intensity of the work in laboratories, whose numbers are constantly increasing all over the world. It has been said that the number of scientists alive today exceeds 90% of all scientists that ever lived on the face of this planet, which is becoming too small to contain them all. They try to spread over the cosmos, invading the moon, and next, Venus, Mars, Jupiter, discovering new galaxies. However, the old question about the opposition between pure and applied science gains new impetus, as if the first (pure science) could yield to the second (applied science) just as all of scientific investigation could concentrate on the discovery of practical means of improving man's living conditions. The pure scientist who is accustomed of doing "science for the sake of science," like the artist who is committed to art for the sake of art, *ars gratia artis*, may even feel ashamed to obtain results which are not designed explicitly to solve some problem of public health, reduce pollution, preserve the environment, economize energy or discover new sources of food.

Paradoxically, the imagination of the common man, *the man of the street*, of all of us who walk in the streets, draws freely upon what science achieved in the past and turns it into "science fiction." Any spectacular show of destruction of what has been built up, by science and technology, earthquakes, burning skyscrapers, retrospects of the two world wars, the Vietnam and Mideast (Middle-East) wars, revolutions, kidnappings, assaults and mass executions attracts public attention much more than an exhibition of any of humanities' great conquests or of constructive themes showing the scientist in action, the artist and the writer striving toward the realization of original work.

It is as if the common man were enraged by what he received almost gratuitously through the work of many generations of thinkers in the fields of sciences and the arts. All the creative activity which has always contributed to alleviate the sufferings of mankind apears to be unable to detain the wave of destruction, which seems to grip us all, from the common citizen to the President of the United States [this was written during the bombings of Cambodia and Vietnam], as well as the international organizations such as NATO, the OAS, the UN and the countries allied in the Warsaw Pact. Everything that was conceived and designed to protect the future of mankind turns into a mainspring for violence and destruction.

Why has this willful anxiety of destruction become so apparent only recently? Two factors may have contributed. The first is obvious and has to do with the diffusion of Western culture; more and more people have access to critical appraisals and the sources of intellectual development. Their numbers keep growing, encompassing many of those who in the West and the East have always been at the margin of civilization. The so-called third world populations begin to partake in determining trends in world history, and they are doing so with an understandable anxiety and violence striving to occupy territories unwillingly and treacherously granted to them. Even in those countries which are parts of what we could call the first world (capitalist) or second world (communist), as standards of living increase for the majority so do aspirations toward more power and sharing of the privileges up to now reserved for the elite and ruling classes. Universities are invaded by a multitude claiming the same quality of education once accessible only to the few. Centers of higher education face increasingly difficult problems in attempting to accommodate a growing and ever more heterogeneous student population. Students, far in excess of those that could be well-educated by existing institutions, find themselves being taught by a limited number of good teachers and unlimited numbers of bad ones. Even if it were possible to produce a sufficient number of good teachers, the job market and the availability of educational resources couldn't expand at anything near a satisfactory rate. It must also be considered that such limited funds are administered by persons not always competent or enlightened who have a greater and greater say in the direction and educational objectives of the universities.

* * *

The result is a general dissatisfaction among teachers and pupils generating conflicts, conflicts which reflect the existing antagonisms between parents and children, employers and employees, politicians and constituencies. Such conflicts or social contradictions do not conform to the marxist scheme of class struggle. Members of the same political or social community turn against each other. (What is happening in Argentina, among other countries, is typical.)

Situations of multiple conflicts have usually been associated with the emergence of a *scapegoat*, and this leads us to the second reason for dissatisfaction, a reason associated with the heightened awareness of problems posed by technological development. Because of abundant discussions in the press, radio and T.V. and by spokesmen for government agencies, the public can easily focus on science as the great culprit of the dissent of our times. Science is ultimately considered responsible for pollution and the destruction of the natural environment, for the indiscriminate annihilation of defenseless populations caught inadvertently in urban and rural guerila warfare, for the use of atomic and nuclear weapons, for napalm, etc. The tragic aspect of the question is that such an attitude is to a large extent justified and in reality the science that originated the industrial methods in chemistry, in physics, in mechanics, in competition for the discovery of fertilizers, and new insecticides, the discovery of atomic and nuclear bombs, might really be the cause, or one of the causes, of such generalized revolt against the *establishment* responsible for such technological advances. Science is indeed serving as a scape goat, not only in countries where it *has* contributed to poison the environment with industrial pollution and automobiles, but also in those other countries, the so-called underdeveloped ones, that manifest a preventive attitude against science, as if the first biscuit or textile factory would usher in a prelude of inevitable bad consequences. In fact, industrial development based on science has given rise to such widespread levels of pollution that no community can feel totally free from its effects. It invades homes, often transported through waterways from distant localities.* Supermarkets, local stores, all distribute pollution in the form of detergents, plastic utensils, containers made of plastic and metal instead of glass; gasoline stations pump pollution into your car which then spreads it over cities, roads and into the countryside. The general clamor against such undesirable effects disregards the fact that they are the mirror image, the shadow which accompanies all the advantages brought about by science and technology.

The obvious solution would be too hazardous. What would happen if we attempted a return to a cleaner rural existence, closing factories, substituting more primitive means of transport such as horse-drawn carts for private automobiles and millions of donkeys for the Sunday drive? Such measures would entail misery and even death for millions whose livelihood is linked to the abuse of pollutants. The gain would be a healthier

---

*As recently seen in a peaceful small town in the interior of São Paulo which was invaded by a mysterious foam. For its inhabitants this represented the first sign of the inevitable pollution.

environment for the few who could survive the absence of pollution. In the Victorian era the populations protested vigorously against the scourge of the locomotive because the burning of coal led to innumerable forest fires and destroyed farm lands. In a very real sense this kind of pollution is *essential* or at least an unavoidable concomitant of industrial development and urban life. There are no straightforward corrective measures for this type of pollution.

In order to ban the widespread use of automobiles, for instance, it would be necessary to develop a whole new collective system of transportation. This would obviously entail radical changes in the lifestyle of people in all of the developed countries (one is reminded of the effects of the acute energy crisis of 1973-1974, in Washington, for instance).

There is, however, the other type of pollution that could be called "de luxe," nonessential, unnecessary and even criminal. An example is the destruction of the environment largely for the purpose of maintaining high profit margins through the disproportionate use of insecticides, herbicides, and fertilizers, when a more equalitarian distribution of lands could give the same gain to the population without the dangers of the systematic destruction of the environment.

From this standpoint, economists who argue against the low productivity of agriculture in the underdeveloped countries, in comparison with the large productivity or profitability of lands saturated with fertilizers and insecticides in the countries with great economic power, are greatly mistaken. In fact, the clamor in such overdeveloped countries, is directed exactly against the abuse of fertilizers to increase by a few percent the profitability of lands owned by the same magnates who influence governments and congressmen against the passage of protective laws. A lucid and still up-to-date explanation of the situation existing in the United States can be found in the book by Barry Commoner, *The Closing Circle: Nature, Man and Technology* (Bantam, New York, 1972).

This is the great problem of the overdeveloped countries, where science and technology led to the growth of megalopolis, and the retreat from the countryside. Thus we have the problem of the large cities, such as New York, Los Angeles, Chicago, and the megalopolis along the Rhine (Cologne, Bonn, Düsseldorf) where pollution from human and chemical wastes attains alarming levels. New York approached bankruptcy, Los Angeles became almost inhabitable (as São Paulo and Mexico City), and the American Lakes (such as Lake Superior) became both poisonous to animal life as a consequence of mercury deposits originating from pesticides, and loaded with algae as a consequence of the accumulation of fertilizers carried down from the overproductive farmlands of Minnesota and Wisconsin. According to a recent report of the World Health Organization,* the large rivers which cross different countries, such as the

---

*Helmer, 1975. "L lucha contra la contaminacion del agua," *Cronica de la OMS*, 1975, 29, 465.

INTRODUCTION 7

Danube, the Rhine, the Mekong or the Niger are those that offer the greatest problems to the international control of pollution arising from chemical products and human waste thrown into their waters. This shows the insensitivity of the powerful toward the misery of their neighbors. The attitude of the mayor of Amity City trying to hide the dangers of the shark in *Jaws* to avoid the collapşe of the summer season through the loss of tourists was seen to symbolize the mayors in Brazil hiding the pollution of their beaches to save profits brought by swimmers on vacation. It was the symbolic implication of the fantastic episode of the shark in the best-seller by Peter Benchley that may explain the impact of the book and the film all over the world.

* * *

But what about science? In what sense the defense of science may help to solve the situation? Even if one recognizes that it has created the problem, science still remains responsible for the improvement in conditions of hygiene, for the raising of average life-expectancy, and improving the means of communication and transportation that have transformed the life of the common individual of the middle class into that of a millionaire of bygone centuries. Bygone centuries? In the beginning of this century only very rich people were able to go perhaps once or twice to Europe and a very few to the United States, Mexico or Japan. A trip to Buenos Aires was a privilege anxiously desired by the middle class individuals and an impossibility to those belonging to the class immediately below the middle one. Only a few happy people could afford the joys of a trip on the large ocean liners which would make stops in Rio or Santos and carry travelers loaded down with luggage, to a sojourn in Buenos Aires (I remember my uncle, one of the owners of Cafe Paulista in Buenos Aires, who was able to cruise back and forth to our great amazement.) It was not so much the price of the tickets that improved (making it accessible to the common man's budget) from the decade 1930-1940 to today, but the means of communication, namely the radio and TV bringing to each one the facilities of the trip, by bus, airplanes or boats. Whereas a trip to Europe (from Brazil) would at one time take a month, nowadays it takes just a few hours to go from Rio to any part of the globe.

This is not intended as tourist propaganda. It is rather propaganda for the science that brought it all into the realm of possibility for so many; not only do we have in mind the practical aspect of science called technology, but also that other science, pure science, that is the basis for the development of applied science. To understand anything man must try to know its mechanism. The scientist starting to work in any investigation must use a mental mechanism which, in its essence, is not very different from that of the child exploring his environment and aquiring the grammar of his mother tongue. The scientist, as the child, can spend all his time rediscovering what others have seen before, but the capacity to invent, to come up with new concepts in science, just as for the child to utter totally new sentences, depends upon a certain audacity, of exploring the "potentialities of the human mind." Just as the child feels in utilizing his mother tongue that there are unacceptable sentences (what linguists call aggrammatical sentences) the scientist also can feel that certain ways are forbidden or will lead nowhere. But the most important analogy can be found in the ability of the scientist, as of the child, to discern that something that has never been thought (or said) can have some generative virtue (transformational, in the sense of modern linguistics), and is the origin of scientific creativity (or of generative grammar).

* * *

The revolution introduced by Noam Chomsky in 1957 establishing a new basis for what has been called psycholinguistics (see Judith Greene, *Psycholinguistics*, 1972) had its highlight in the publication of his book, *Cartesian Linguistics* (cited in Chapter III) and in subsequent books: *Aspects of Synthatic Structure* (1967) and *Semantic Questions* (in French, *Questions de Semantique*, 1975). Not everything that was said by Chomsky can be accepted without discussion (for a strong but constructive criticism, see I. Robinson, 1976). Some of his contentions (and opinions) have strong affective or merely intuitive character, but it is doubtless that his definitions of grammar as comparable to a theory in science, the transformational analysis of sentences, superficial and deep structures (reviving some concepts of the XVII century: grammar of Port Royal, Cartesians and so forth) leading to the idea of the existence of a universal language interiorized by the individual who learns his mother tongue, and the concepts of grammaticality, linguistic competence which contributed to revive Cartesian and Humboldtian concepts of language as the basis for thought, as well as the rational use of Humboldt's slogan that "grammar uses finite means to attain infinite objectives," and so forth. All that became the object of hot discussions among linguists and philosophers (see Seuren, 1974; Katz, 1964; Robinson, 1976 and many others) could not leave indifferent those interested in understanding the way things are learned (in a general way), not only in language but also in science.* Cognitivism does not obviously exclude the value of the contribution of the behaviorists of Skinner's school, who intentionally confine themselves to the objective conditions of "emission of behavior" in laboratory situations. Between them there exists the same difference that exists between the visionary designer of a cathedral and the practical architect or the bricklayer who

---

*In a more recent book, *Reflections on Language* (1975, 1977), Chomsky treats the subject in a more esoteric way.

executes the plans. It is the eternal difference between the theoretical man of science and the technician who establishes conditions to verify theories proposed by the former.

We all know that mathematical knowledge is not the only knowledge that depends upon pre-formed patterns in the human mind. The functions of the human mind are always those of choosing what *seems* to be true among the multiplicity of facets that natural occuring phenomena may assume. This choice is far from being arbitrary because it must be submitted to a counterproof or a control of the phenomenon itself. At the same time, the experimental proof may suggest substantial alterations of the theoretical model. One may say, in cybernetic language, that the phenomenon as observed is the *feedback* of the theory. Theory, counterproof and revision of the theory form a totality which is the essence of the *scientific method*, and cannot be ignored by anyone wishing to advance his science. But there is a moment in the original process of creation when the method has not yet been fully applied and when the scientist intends the formulation of a new scientific truth. At that moment the scientist resembles a child acquiring his mother tongue. The criteria he uses to select (or discriminate) is analogous to that of the child who, between two sentences of his mother tongue competently chooses the one that is grammatically possible according to the "universals" of human language. The sounds (phonetics) associated with the meaning (semantics) are epiphenomena dependent upon environmental or racial conditions which determine the "superficial structure" of the uttered phrase. In different languages the meaning of the phrases: "Tenho fome," "I am hungry," "j'ai faim," "Ich bin hungrig," and so forth, represents the universal semantic in the corresponding languages.

However, the problem of associating semantics (meaning) to phonetics is of extreme complexity, and has been only partially disentangled by transformational grammarians since Katz and Fodor (1963) and Katz and Postal (1964) that for the first time tried to solve the problem propösing a provisional "lexical" solution. A lucid discussion of the problem is presented by Geoffrey Leech (1974) in his book *Semantics* (published by Pelican Books, Middlesex, England), and a very difficult one by Chomsky (1975-1977).

* * *

One never ceases to be amazed at the intuition of the young Einstein who at age sixteen formulated a paradox that later became the basis of his theory of relativity (restrict relativity). If one were traveling with the velocity of light, following a light beam closely, one would perceive the latter as losing one of its characteristics, that of obeying Maxwell's law. The light beam would appear to the observer moving at the velocity of light as an electromagnetic field "at rest." In accordance with Einstein himself who described this paradox in his memoirs, "the beam of light should appear to the observer as if the latter were at rest, because for him there would be no way of deciding whether he would be in uniform motion, with whatever velocity."* This paradox was solved a few years later, when, in 1905, at the age of twenty-five, Einstein proposed his theory of restrict relativity, which stated as a fundamental postulate, that no observer could travel with the velocity of light, and that Maxwell laws are invariant in relation to the movement of the observer.

The practical importance of this truth, chosen among many possibilities, we all know: the deduction that the mass (m) is equivalent to an energy (E) which may be calculated by the famous equation $E = mc^2$, generating the atomic and nuclear bombs, the pacific or non-pacific use of nuclear energy, Hiroshima, Nagasaki, and the foundations of the cold war for the conquest of the secrets of the atom. That among millions of educated people reaching sixteen years of age, only one had such an intuition, continues to be the great mystery of *scientific creativity*.

The problem of learning science or scientific creativity received new inspiration through the considerations of the Cartesian linguists of Chomsky's school. One may find oneself turning around in circles trying to establish whether there is something mystical in scientific thought or something scientific in mysticism. All the questions raised by significant philosophers of all times (Pythagoras, Plato, Aristotle, Locke, Hume, Descartes, Leibnitz, Hegel, Bergson) and by scientists who are also philosophers (Descartes, Gallileo, Poincare, Mach, Einstein, Planck, Bohr, Schrödinger, Heisenberg) and also by biologists (Claude Bernard, Pasteur, Darwin and many others) have had to do with whether scientific creation is exclusively objective; i.e., totally dependent upon experimental observation or if the mind contributes with some sort of "mental blueprint" to perceptions. In the ceaseless search for such a blueprint, the errors have been frequent, from the *Idea* of Plato, to the dogmatism of Kant, considering space and time, as "*a priori* synthetic judgments," i.e., as structures confined strictly to the mind to render possible the ordination of the observed phenomena. Nowadays Kant's mistake has become obvious because we know that "real space" may not be that isotropic, homogeneous and absolute and exist as a receptacle to natural phenomena. It appears that "real time" is indissolubly connected (bound) to "real space" as a four dimensional continuum, therefore the world in which we live must contribute with something local and specific for the representation that we should make of space and time; in other words in Kant's hermetic nomenclature space and time would be analytical judgments combining something of intuition with the experimental data. And here we find one of

*Hoffman, "An Einsten Paradox," *Transactions, N.Y. Acad. Sciences*, 36, 721, 1974.

the tricks of nature, imposing obstacles that must be removed in the pathway to truth.

The extension of our ignorance can be evaluated by this brief citation: 'We don't have yet a quantic theory of gravitation that might be considered satisfactory. We do not yet know if a new type of treatment distinct from that of Newton-Maxwell-Einstein quantum mechanics would be required to understand the subnuclear phenomenology. Maybe the notions of space and time though necessary as a scenery of our knowledge of the physical world (macroscopic, atomic and nuclear) would be unapplicable to domains below the extension of 10^{-13}cm or to high energies. Which new ideas would then take the place of Kantian *a priori* forms of space and time of our pure intuition?" See J. Leite Lopes, "L'Evolution des notions d'espace et de temps." *Scientia*, May-July, 1972.

Chapter I

SCIENCE AND THE MAN

I
SCIENCE AND THE MAN

The fascination that modern science exerts upon the common man, the layman, has several causes, among which we could mention the following:

1. The marvelous capacity of technology (usually confused with science) to improve public health, to increase the production of food and domestic utilities (plastics, home appliances such as TV, radio, phonographs, and so forth), automobiles, airplanes, and everything that exists today and did not exist fifty years ago, when science and technology began to join efforts for the great industrial revolution of the century. Combined with a sense of bewilderment there is a contradictory feeling of satisfaction and fear, since the ability of science to generate new tools for everyday life is also associated with its capacity, no less amazing or miraculous, to generate instruments of destruction, such as atomic and nuclear bombs, defolients, toxins and venoms that can be used in bacteriological or chemical warfare.

2. The great capacity of science to forecast future events with a relatively small margin of error; and here we may mention the admiration which the layman has (we are all laymen ourselves, even the most specialized among us who nevertheless ignore extensive domains of human knowledge) for the ability to make astronomical predictions which in ancient times were part of the mysterious science of sages and prophets. Nowadays such predictions are contained in the almanacs for agriculture, commerce, transportation, and so on, but our admiration is often curtailed by failures in forecasting. As witness to these we have the scandal of the predictions of the Kohoutek Comet which had been expected to appear at a certain time (January 1974) but failed to do so in the bombastic way predicted by astronomers and physicists. Similarly, the small inaccuracies in weather forecasts may give laymen a bad impression of the efficacy of computers in manipulating data or "parameters" used for such predictions; meteorologists have the tendency to defend their methods by calling upon the

complexity of causative factors and of the processing of data by computers in order to justify the imprecisions in forecasting snowstorms, tornados, and earthquakes, or simply the movements of a cold front or heat wave. (The situation has now improved to a certain extent with the images received from satellites, but still, misreadings of such images are not infrequent.) The same could be said about the use of computers in hospitals for predictions about treatment, diagnosis and prognosis of certain ailments, or in the stock exchange market, in government executive offices where errors range from minimal deviations from expected results, to complete failures in the forecasting provided by computers. It is true that the mistrust in the efficacy of scientific methods can largely be compensated by the great achievements of astronauts or cosmonauts successfully placed into terrestrial, lunar, marsian, venusian, mercurial or solar orbits, and controlled from a distance of millions or billions of miles. Watching on TV the fantastic show of views from the surface of Mercury, or the craters of Mars, is a positive gain against the failures of hurried astronomers announcing the arrival of the "Comet of the Century" (superior to Halley's Comet) which turns out to consist of an almost invisible luminosity comparable to that of a star of third or fourth magnitude.

3. To anyone who has had an elaborate scientific training (such as is required of candidates to Brazilian universities) science can show an intrinsic beauty justifying the total dedication of some to the elaboration of theories or new concepts about the essence of the universe. Very few can fail to admire the imposing architecture of modern physics, chemistry, and biology. For many, the most sophisticated ones, scientific invention has something to do with the re-creation of the Universe, as if the Creator had used all the resources of quantum and wave mechanics, the *Aufbau* principle, the genetic code, to lay out the plans of atoms and molecules creating inter- and intra-atomic and nuclear forces, the movement of electrons in their orbits, synthesizing the ultimate constituents of living matter, according to the principles of transmission of information in the living body, enzymology and genetics, and everything that constitutes the foundations of the knowledge to be acquired by the candidate to schools of engineering, chemistry, medicine, pharmacy, and dentistry. This fabulous mass of knowledge that must be swallowed and badly digested by any candidate for a modest place at the university (in Brazil) presents itself as the superessence of logical and creative activity of the human mind assisted more and more by complicated equipment and computers which begin to be incorporated into the everyday lives of high school or college students, and many believe that soon they will be part of the environment of first-graders and even kindergarteners, who after all are the ones who will come of age in the twenty-first century. But, again and again, one is reminded of those who scorn the grandiose architecture of modern science, sometimes

called Western science, in favor of certain mystical tendencies of Oriental heritage: Chinese, Hindus, Persian, Muslim, in the interpretation of natural phenomena. Such myths and thoughts constitute the bases of philosophies still dominant in several regions of the globe. Here we find the real sage, the contemplative man, in opposition to the scientist, as the one who sees and hears truths emanating from an invisible world which can be perceived only by the initiated, or simply by those with mediumnic powers (Carlos Castaneda calls Don Juan a *man of knowledge*). If such truths are in conflict with the doctrine established by the scientist, it does not mean that the former are false and the latter true, but simply that the latter (the scientific) represent a simplification or frozen view of such eternal truths that can only be obtained by the other method, the mystical or magical one of the real sage. (*See* Note I at the conclusion of this chapter.)

This situation was described in an attractive way by the anthropologist Carlos Castaneda, a sorcerer's apprentice, in his book *Teachings of Don Juan* (1974), and in a more synthetic way in *Separate Reality* (1976). In these books, the vision of the world learned at the University of California in Los Angeles was subjected to wild criticism by the alleged sorcerer (bruxo) Don Juan of Sonora. (*See* Note Ia.)

This tendency to interpret the universe in two different ways has been present throughout the history of mankind, and certainly there was a time when revelation was predominant as the ultimate manner of discovering the harmony existing or inherent in the laws of the Universe. Only very recently in the last three or four centuries the tendency to impose the other, the scientific way of interpreting natural phenomena has cropped up slowly but is not yet completely established. This explains why there are the oldest civilizations, Chinese, Hindu, Persian or Muslim that cultivate in a more strict way such kinds of natural philosophy, mystic, magic or mediumistic. We may say that the Mexican sorcerer (bruxo) uses an experimental method (medicinal herbs, hallucinogenic plants, peyote, mushrooms, solanaceae) to expand his vision of the world, to "see" instead of merely "look" at things (in a similar way as scientists and physicians use LSD as adjuvants to psychotherapy or diagnosis). (*See* Castaneda, *Separate Reality*, 1976, Note Ia.)

We must not ignore that within Western culture, itself rooted in the European Renaissance, many scientists (and not the minor ones) would blend their scientific discussions with mystical and magical interpretations. One of these and perhaps the most famous, the astronomer Johann Kepler, aside from his fabulous discoveries of the laws of movements of the planets, would assume a not small dose of mystical interpretation to explain what he called *Harmonice Mundi*. Isaac Newton himself after achieving the greatest scientific synthesis known up to his time (or in the history of science) maintained his famous *scholium* in which he would place God as the guardian of the harmony of the universe, a kind of policeman to guarantee the execution of his laws of universal gravitation, and used to consider the most important part of his contribution the commentaries of the Holy Scripture, a task to which he dedicated himself at the end of his life. We will find among the most famous biologists those who accepted the ideas of vital force, spontaneous generation, and would interpret the organization of living matter as an *entelechia* in the Aristotelian sense of the word. And even today not all scientists who occupy themselves with the functioning of the central nervous system (CNS) would be prone to reject the interference of supernatural powers to maintain the duality "soul-body" just as vividly as the scholastic philosophers or the learned men of the Renaissance.

It is a common fact that for the scientist the participation of such influences is assumed (when it is) as a mere hypothesis to be eliminated as our objective knowledge of the functioning of the living organism progresses. They have always represented a "no-man's land" or vacant lot over which the science of the future will be erected upon sounder foundations. But while this does not happen, the sage, the mystics and the magicians use those unexplored territories for their fantastic constructions which to the scientist are no more durable than the tin huts of the shanty towns (*favelas*), which grow in void areas in the heart of the city and tend to disappear whenever urban designers lay claim to the same land.

The main argument of the scientist against such provisional constructions has always been that the use of such concepts, if valid at a subjective or abstract level, have never contributed to improve human life. It has never been possible to employ mental energy, vital force, *nirvana* to build up a motor or a machine that might improve working conditions.

Vital force, if it existed independently of the physical or chemical energy contained in the molecular interconnections of food, would enable the construction of a moto-continuum making us independent of food consumption. The simple fact that oxen driving a mill must receive their food ration at regular intervals and that laborers need double or triple amounts of caloric intake compared to the sedentary clerk, are arguments against the existence of anything supernatural in the so-called vital force as understood by mystics and vitalists of all ages.

The balance of basal metabolism in which everything can be calculated in terms of calories, ingested, stored, or eliminated, is a strong argument against the vitalistic interpretation of the functioning of the biological machine. The same could be said of such a concept as mental energy, which is still hypothetical, and the impossibility of utilizing it to move bicycles or to serve as supersonic aircraft fuel.

But obviously, the purpose of mystics, magicians and medium is not to construct terrestrial or flying machines to compete with those fabricated by human technology. What they pretend at most is to explain human behavior in its most subjective aspects; i.e., the intentions behind human behavior. And here we find that what we have called no-man's-land or identified as vacant lots can be immeasurably extended to accommodate constructions of all kinds, from the house built with reinforced concrete of the behaviorists to the provisional huts of the mystics and magicians.

What enables such provisional construction still to have a certain force is the enormous ignorance of the scientist about the functioning of the central nervous system (CNS), and of the sense organs. Whatever the scientist is able to discover regarding the functioning of the CNS can be used by the mystic-savants to strengthen new provisional constructions. An example is the observation of alpha-waves in the electroencephalogram of the normal individual when totally at rest, in silence, and with eyes closed.

The existence of such waves has been used by many scientists and mystic-savants for the development of more or less complicated instruments capable of interfering with the functioning of the human brain. Today it is common to treat anxious or disturbed patients by interfering with alpha-waves, which means no more than to apply electrical oscillations of a slow rhythm and high voltage to the skull.

At the same time science is developing *biocybernetics,* in which well-documented knowledge of electrophysiology is being mixed up with dubious concepts of mental energy as well as of the influence of brain stimuli upon the normal or pathological functioning of the organism. All this represents the battle raging in no-man's-land where solid buildings share the space with the miserable shanty towns of the marginals of science.

4. Finally, science tries to invade fields that have always been under the domain of erudites and mystics, or else of men of letters and novelists. Sociology became a science under many aspects provisional, but where the methods of objective (natural) science are applied in a serious way by researchers with a solid scientific formation, though many of them did not succeed to overcome ideological dogmas about the natural event to be interpreted. This aspect was treated in a recent publication.* Psychology becomes more and more a respectable science gaining ground in such no-man's-land in which the methods of natural science are building up solid constructs by the behaviorists, in opposition to the subjective and dubious methods of the gestaltists, psychoanalysts, and psychologists of the old school (introspectionists, functionalists, and so forth). Finally, linguistics undergoes a profound renewal becoming rigorously scientific, as in the study of phonetics or phonology (by using electronic means of registering sounds, etc.), although in some of its sectors (syntactics and semantics) it is still very difficult to separate the wheat from the chaff; i.e., to distinguish

*M. Rocha e Silva: 1976, Ciência, tecnologia e educação como base do desenvolvimento. Reuniâo UDAL, México, Nov., 1976).

what could be considered a newly discovered scientifically grounded fact from what is nothing more than a mental construct, a *vue d'esprit*, a hunch or simply a working hypothesis. (*See* citations in Chapters III and IV.)

In a discussion on the sources of irrationality, Frankel (1973)* analyzes the irrationalist attitude of philosophers and modern concretists, coming to the following propositions:

> 1) The universe of man is divided into two realms—one of appearance, the other of reality. The former is marked by accident, doubt, uncertainty, frigidness, alienation. In the second, doubt is dispelled, time and death have no sting.
> 2) The reason why people mistake appearance for reality is that their definitions of reality rest on biased presuppositions which their culture, class and practical concerns impose upon them.
> 3) Human nature exhibits this ontological dualism between appearance and reality. A war goes on inside each person between the "cerebral" and the "emotional," the "conscious" and the "intuitive," the "empirical" and the "rhapsodic."
> 4) The unmistakable sign that we have gone astray is when we arrive at states of consciousness in which subject and object are distinguishable. Then science becomes untrustworthy, on principle, because it rests on the distinction between the objective and the subjective.
> 5) Accordingly, all problems, cognitive, emotional, and social are reducible to a loss of harmony between man and his environment, his head and his heart, his ideas and his instincts. Thus beyond its assertions about the nature of man and the universe, irrationalism offers an image of the good life. It is a life free of unrest and unease—a life released, through passionate ecstasy or rapt contemplation, from the regretfulness of time, the vexations of decision, the risks of fallibility. It offers the vision of a kind of peace and unequivocal acceptance and commitment from which the normal perils, pains, and worries of human existence have been removed."

The great question that must be answered is whether modern science could accept such a dualism as an object of investigation. There is no need to argue that such a question has been formulated and occasionally even answered by philosophers of all ages, and especially by psychologists, and anthropologists, behaviorists, mentalists, structuralists, functionalists, or whatever name may have been given to doctrinaire followers of Freud, Jung, Skinner, Chomsky. Their goal is to stand in trenches considered inexpugnable, and to a certain extent immune to adverse criticism even from those who, not being experts or adversaries, understand the eternal dualism latent or explicit in the great myths created by men to express religious and scientific conflicts over the natural order of things. The taking of positions by the experts of this or that doctrine usually reflects peculiarities of educational background. One might give some examples.

*C. Frankel, 1973. *Science*, 180, 927-931.

CHAPTER I

In the Anglo-Saxon countries in which reading the Bible is an almost universal occupation, there is a tendency to interpret visions and oneiric images on the basis of the great biblical myths. These are based on Genesis, the expulsion of Adam and Eve from earthly paradise, the ladder of Jacob, the Babylonian temples, the birth, life and death of Jesus Christ, and are seen as representing what is deeply routed in the subconscious of the species under the form of archetypes, mandalas, cherubs, St. Georges and dragons, hell, heaven, and purgatory. For other people, however, such symbols and situations may be nothing more than objects of artistic creation, as in the Divine Comedy, in the Decammeron, in the pictorial works of Fra Angelico, Boticelli, Leonardo or Michelangelo, El Greco, and all of those whose work can be seen in the museums of the Vatican, the Uffizzi, the Louvre, and el Prado, where the best religious art from the middle ages and the Renaissance has been preserved.

In such masterpieces of human genius the archetypes become conscious through forms dominated by loftly aesthetic values. Not everyone is capable of creating or even truly perceiving such forms. For some the archetypes remain embedded in their scriptural literal sense. Having been transmitted throughout the ages by means of cultural traditions, they gained a grip upon the unconscious mind. To claim that the transmission from biblical times to our modern era of supersonic aircrafts and atomic bombs may have been genetic is as ludicrous as the discarded notion of inheritance of acquired characters.

It is widely accepted even by the creator of psychoanalysis himself that his ideas were greatly influenced by educational and life-style patterns typical within certain social classes in Europe during the late-nineteenth and early-twentieth centuries. The public revolt of Jung against the master arose mainly from personal divergences in the interpretation of psychoanalytic dogma, divergences determined by the intellectual and moral background of Jung, in contrast to that of the master. If Jung's interpretation became more acceptable it is due to the recognition in Freud's method of certain peculiarities which appeared too subjective or personal. This great crisis within psychoanalysis has been vividly described by Fromm (1971), in the *Autobiography of Freud* (1935), in the writings of Jung (*see* Notes II, III and V) and represented the schism that shook the prestige of the whole movement during the critical phase of its consolodiation all over the world, but especially in Europe. (*See* Notes II and III.)

When the crisis is considered from the viewpoint of the ethical norms of Western science, it represented a scandal hitherto unheard of in any sector of natural science. When Einstein with his relativity theory had shown the insuffiencies of the Newtonian theory of universal gravitation there was no attempt of personal retaliation from the adepts of Newtonian theory; or when Planck announced his theory of the quantic emission of radiation if there was any resistance or suspicion about the acceptance of his basic postulates of quanta theory, in substitution to the basic postulates of equipartition of energy along the wave lengths of the emitted radiation, scientists involved in the conflict were to a certain extent protected by what is commonly accepted as the basis of scientific ethics.

It is a fact that doubt or suspicion on the part of reputable elements of the scientific community may lead a great scientist even to commit suicide (as was the case with Ludwig Boltzmann, the founder of statistical mechanics) but even in such extreme cases, the conflict tends to remain within the limits of questions of principle, without affecting the personality of the scientist involved.

In the case of the schism in the domains of psychoanalysis, the conflict assumed a strong personal character as can be seen in the quote from Fromm. (*See* Note IV.)

The deepest conflict in the psychoanalytical movement has been the dramatic split of Jung from his master. To Freud, the unconscious was inhabited by shadows and ghosts repressed from the conscious mind by an autotherapeutical procedure by which what could be dangerous to the individual would suffer a process of censorship through the consciousness derived from external pressures of the social environment of the individual. On the contrary, what Jung called *unconscious* was the great ocean in which the *ego* floats almost totally submerged as an iceberg. During the first years of an individual's life, up to the age of ten or twelve, behavior is almost exclusively the product of forces that originate in the unconscious and that slowly consolidate the individual ego which becomes more and more conscious.*

We have sketched here the conflict between behaviorists and mentalists, because to Freud the composition of the individual ego derives from external influences in a certain way continuing the line of the English empiricists, who in turn were affiliated to the Aristotelian concept: *nisi est in intelectu quod primum non fuerit in sensu,* reformulated by Hobbes and Locke in the seventeenth and eighteenth centuries, and known as the *tabula rasa* model of the human mind (*see* next chapter). To Jung, the champion of analytic psychology (*see* Notes II, III and IV), the analysis of dreams would lead to the identification of archetypes; i.e., of what is most primitive in the human soul. Such archetypes or archaic patterns or concepts find their way to consciousness provided the barriers from the conscious are lowered or even destroyed during sleep. According to Freud (*Interpretation of Dreams*) the substance of the dream originates from the external world and must be found first in the material that went through the conscious mind and has been repressed to the unconscious (or subconscious). This appears to be the main or dominant function of the

*C.G. Jung. 1973. *Approaching the Unconscious. In Man and His Symbols.* A Laurel Edition.

superego that, in Jung, is part of the "internal organization" of the unconscious maintaining its content within the limits imposed by the evolution of the species. (*See* Note V.)

A similar situation can be found in the recent discussion between Skinnearians and Chomskysts, (*See* J. Lyons, *Chomsky,* Ed. Seghers, Paris, 1971.) or better, Chomsky's attack of Skinner which assumes that a certain politician stance lies behind his behaviorist theory. Skinner does indeed propose a society controlled by a "technology of behavior," which he describes in two books, *Walden II,* and *Beyond Freedom and Dignity* (Bantam/Vintage Books, New York, 1972). Possibly stimulated by his extraordinary success in the field of linguistics, the history of which can be divided in two periods (BC and DC, before and after Chomsky), and on the basis of his liberal ideas first made public during the anti-Vietnam war movement, Chomsky expressed his ire against the Harvard psychologist in a series of articles dating from 1959 (Chomsky, *Review of Skinner's Verbal Behavior*, 1959) to 1974 (Chomsky, *For Reasons of State,* 1974) culminating in a catilinary published in the New York *Times,* included in his publication of 1974, in which he criticized the linguistic conceptions of Skinner* in connection with his political tendencies which, according to Chomsky, have been made very explicit in Skinner's most recent writings. The latter considers himself dangerously "above dignity and freedom" (B.F. Skinner, *op. cit.,* 1972.)

What can be seen through such an antagonism in many respects very stimulating is the taking of positions of two distinguished men of science, one for concretism (Skinner) and the other for subjective mentalism (Chomsky). The political and personal arguments merely constitute strategic attitudes to the final assault to the frontiers defended by both contenders. Such a dualism is once again a manifestation of that dualism between the empirical and rhapsodical that will constitute the main subject of the essays that follow. In previous essays (M. Rocha e Silva, *A Evolução do Pensamento Cientifico,* Hucitec Edit., Sao Paulo, 1972) we have concluded that such conflicts may be considered as being different aspects of a complementarity defined by Bohr (N. Bohr, *Physique Atomique et Connaisance,* Ed. Gouthier, Paris, 1961) that exists between human cultural manifestations. From such a viewpoint such conflicts can really help to improve the understanding of the objects of science, either pure or applied.

In the following chapter we will analyze some aspects of the endless struggle between the abstract and the concrete. The philosophies of all times have been marked by the opposition between abstract and concrete, analytic and synthetic, soul and body, conscious and unconscious, essence and phenomena, form and matter, finite and infinite, freedom and neces-

---

*Skinner, B.F. 1957. *Verbal Behavior*, Appleton Century Crofts, N.Y.

sity, possible and real, theory and practice. The examples mentioned in this chapter show that mankind continues to struggle between two poles that could be expressed in terms of another opposition: tragedy and comedy, the first resulting from a subjective interpretation of different ideologies, and the second in the Balzachian sense of *Human Comedy,* as a novelist's description of the common man who places in science his hopes for liberation.

Note I

There has been for some time a tendency to criticize the scientific positivism that has dominated science since the English empiricists, the French encyclopedists of the seventeenth and eighteenth centuries, and has continued in our century with the so-called Vienna Circle with so many outstanding philosophers of science of the first third of our century. We can even detect a tendency to return to Kantian idealism on the basis of the existence of a cognitive mechanism in human mind (pure reason) supposed to bear the innate or transcendental patterns for human knowledge, the so-called *a priori* synthetic judgements (space, time and causality). A tendency to mysticism or purely metaphysical or religious tendencies, can be noted among those that consider themselves the *avant garde,* hippies, artists of the new wave (bossa nova); i.e., those continuing the movement initiated with cubism, futurism, existentialism, that direct their fire of the absurd either in a spontaneous, sincere movement, or as a simple instrument to try to stop the dominant bourgeousie (*pour épater le bourgeois*). Evidently, such an obvious subversive tendency could not fail to meet opponents among the reactionaries or simply conservatives of the *establishment*. Finally, such tendencies could not fail to have repercussions upon scientific creation, especially in human sciences.

Note Ia

In the disturbing books by Castaneda, especially in the most synthetic one: *Separate Reality,* it is difficult to decide whether we are in the realm of reality or in the wonderland of a fairy tale. Nonetheless, the distinction between "seeing" and merely "looking at" is a real concern to scientists when they inspect experimental results or thoughts. To *see* is to go deep into the meaning of a phenomenon. It is the attitude that leads to the creation of a new theory that may change one's outlook of the universe, though to *look at* means to browse over the appearances of the observed events. Whether Don Juan or Castaneda are referring to another world which actually exists or to a new world to be discovered behind reality is a matter of literature or of science fiction. In fact, not infrequently the conditions of scientific discovery (see the book edited by Klemm, 1977) are about the same (or similar) as those described by Castaneda in his *Teachings of Don Juan of Sonora*. The scientist, as a "bruxo" (sorcerer, magician) can also "see" a new world in which he starts to live, at the beginning, alone, and usually mistrusted by his colleagues and by laymen. Most scientists merely "look at" things and pick up data of observation. It is also understandable that eating peyote or smoking mushrooms may contribute to open new frontiers of unbelievable beauty or strangeness as described by Huxley *(The Doors of Perception)* and can be found in the inexhaustible data contained in books and reports about opium smokers, LSD users and those who take "trips" via hallucinogenic plants. Therefore the appeal of Castaneda's books can be a real treat to a modern man. I would advise every-not-too-busy scientist to browse over his books, and apply his teachings to his everyday life as researcher.

Note II

The divergence between Jung and Freud has been described in a delicious way in the series of conferences that Jung gave in England in 1935 which have recently been reprinted under the title *Analytical Psychology: Its Theory and Practice*, (C.G. Jung, Vintage Books, New York, 1970). "If you want me to elucidate the question of the connection with Freud, I am glad to do it. I started out entirely on Freud's lines. I was even considered to be his best disciple. I was on excellent terms with him until I had the idea that certain things are symbolical. Freud would not agree to this, and he identified his method with the theory and the theory with the method. This is impossible, you cannot identify a method with science... I know that what Freud says agrees with many people, and I assume that these people have exactly the kind of psychology that he describes. Adler, who has entirely different views, also has a large following, and I am convinced that many people have an Adlerian psychology. I too have a following—and it consists presumably of people who have my psychology... If the whole world disagrees with me it is perfectly indifferent to me. I have a perfectly good place in Switzerland, I enjoy myself and if nobody enjoys my books I enjoy them... I cannot say I have a Freudian psychology because I never had such difficulties in relation to desires. As a boy I lived in the country and took things very naturally, and the natural and unnatural things of which Freud speaks were not interesting to me. To talk of an incest complex just bores me to tears... we do not want to change anything. The world is good as it is."

Note III

"I cannot say where I could find common ground with Freud when he calls a certain part of the unconscious the *id*. Why give it such a funny name? It is the unconscious and that is something we do not know. Why call it the *id?* Of course the difference of temperament produces a different outlook. I never could bring myself to be so frightfully interested in these sex cases. They do exist, there are people with a neurotic sex life and you have to talk sex stuff with them until they get sick of it and you get out of that boredom... It is neurotic material and no normal person talks of it for any length of time... Primitives are very reticent about them... Sexual things are taboo to them, as they are to us if we are natural. Many people make unnecessary difficulties about sex when their actual troubles are of quite different nature." (Jung, *Analytical Psychology,* pp. 143-144.)

Note IV

E. Fromm, *The Crisis of Psycholoanalysis,* Fawcett Premier Books New York, 1971.

"Other disciples went away. Jung, for other reasons because he was a conservative romantic, and Adler because he was a more romantic rationalist, though very well doted... Rank developed original points of view and was sent away, lesser by the dogmatic attitudes of Freud, than because of the envy of his competitors. Ferenczi, perhaps the most amiable and imaginative disciple of Freud... was however harshly eliminated when he deviated from important points at the end of his life. William Reich was eliminated from the organization... and this was an interesting example of the psychoanalytic bureaucracy and also in such a case, by influence of Freud who refused to get away from the reform, to a radical position in the center of the circle of his system."

Note V

About the Collective Unconscious

"The deepest we can reach in our exploration of the unconscious mind is the layer where man is no longer a distinct individual, but where his mind widens out and merges into the mind of mankind—not the conscious mind, but the unconscious mind of mankind, where we are all the same... On this collective level we are no longer separate individuals, we are all one. You can understand this when you study the psychology of primitives. The outstanding fact about the primitive mentality is lack of distinctiveness between individuals, this onenesse of the subject with the object, this *participation mystique,* as Levy-Bruhl (*How Natives think)* terms it." (Jung, *Analytical Psychology,* p. 46.)

CHAPTER I

"Our mind has his history, just as our body has its history... Our unconscious mind, like our body, is a storehouse of relics and memories of the past. A study of the structure of the unconscious collective mind would reveal the same discoveries as you make in comparative anatomy... There is nothing mystical about the collective unconscious... The brain is born with a finished structure, it will work in a modern way, but this brain has its history. It has been built up in the course of millions of years and represents a history of which it is the result. Naturally it carries with it the traces of that history exactly like the body, and if you grope down into the basic structure of the mind you naturally find traces of the archaic mind."
(Jung, *Analytical Psychology*, p. 44-45.)

"The word 'unconscious' " is not Freud's invention. It was known in German philosophy long before, by Kant and Leibnitz and others, and each of them gives his definition of that term. I am perfectly well aware that there are many different conceptions of the unconscious, and what I was trying humbly to do was to say what I think about it... Freud is seeing the mental processes as static, while I speak in terms of dynamic and relationship. To me all is relative. There is nothing definitely unconscious; it is only not present to the conscious mind under a certain light... The only exception I make is the mythological pattern, which is profoundly unconscious, as I can prove by the facts."
(Jung, *Analytical Psychology*, pp. 68-69).

Chapter II

DIALECTICS AND THE SCIENTIFIC METHOD

II
DIALECTICS AND THE SCIENTIFIC METHOD

The abstract-concrete duality appears clearly in the two currents of thought that dominated the seventeenth and eighteenth centuries. The English empiricists were represented by Hobbes and Locke with the basic slogan that nothing exists in the mind that would not proceed from the senses. This in turn could be affiliated to the idea attributed to Aristotle "that everything in the intellect was first (*primum fuerit*) in the senses except the intellect itself," and the other current of thought led by Berkeley and other mentalists of the same epoch. Hobbes' and Locke's conception became popular with the connotation of *tabula rasa* denoting the intellect itself before receiving the imprints of the external world, as a blackboard before receiving the sketches of the pupil or the deductions of the teacher. Already Plato in his *Republic* assumed a variant to this situation in his famous parable of the cave where man, the human species, sitting on the place of the observer with his back turned to the outside world, and therefore unable to receive direct impressions from the events occurring in it, was limited to observation of such events under the form of shadows projected at the walls of the cave. Such shadows though determined by real events by the "objective" or concrete, could only be representations distorted by the mind of the observer. Such shadows would constitute the raw material to build up the *idea* that finally could be the best representation in the human mind of things which happen in the universe.

Therefore, if we affiliate the empiricists to the ideas of Aristotle we could affiliate the mentalists or solipsists of the seventeenth and eighteenth centuries, under the leadership of Berkeley, to the ideas of Plato who saw the possibility to reconstruct concrete events from such possibly distorted images which constituted the only representation of the human mind about the real phenomena.

Such a duality can be detected also among the great philosophers and scientists of the seventeenth and eighteenth centuries, namely those who saw in the representation of the external world the data derived from immediate observation: the concretists or materialists Linneaus, Malpighi, Leeuwenhoek, Cuvier in opposition to those who tried to explain the Universe by means of symbols or abstract thoughts of mathematics, such as Descartes, Galileo and Newton.

On philosophical grounds, the solipsism of Berkeley was up to a certain extent reformulated into a monumental theory of evolution and of universal history in Hegel's conception which endeavored to explain the great events of universal history by the interplay of contradictions existing in the human mind ("Phenomenology of the spirit"). Practically all manifestations of the human spirit through the universal conscience (or absolute conscience) would constitute the dominating factor of determining the course of history, the arts, science and religion.

What distinguished the solipsism of Berkely from the dialectical logic of Hegel or his "phenomenology of the spirit" was that the former (solipsism) constituted a more or less static phenomenon, the representation of the external world as reflected in a mirror, or inversely, the images of the mirror reflecting themselves in the external world, as it was for Alice in the world of the looking glasses. Hegel's dialectic on the other hand endeavored to project into the concrete world of natural events the contradictions which are inherent to pure reason, infinite spirit, universal consciousness, or whatever names Hegel found to denote the motor power behind history. Reason or human mind (or spirit) can be identified with its negation, and from the transformation of the *being* into the *non-being* would be derived all the forces of history, as an external manifestation of the vital phenomenon, contradictory and evolutionary, developing inside the individual himself.

What such a condition might mean, only very few could understand, and it was necessary for the genius of Karl Marx to invert Hegel's formula establishing as a fundamental postulate of human evolution in history the contradictions created by the forces of production that became the fundamental postulate of the "historical materialism" of Marx and Engels. (*See* Note I.)

If Marxism is understood as the predominancy of concrete factors in human production, upon the evolution of the spirit or universal conscience, in a certain sense could be called the *non-Hegel*. Nonetheless, in accordance with his dialectical logic, Hegel himself might be identified with his own negation (the *non-Hegel*), and therefore everything would be harmonized by either accepting Hegel's conception of historical evolution as a consequence of the contradictions existing in the human mind, or by assuming an anti-Hegelian attitude of the historical materialism of Marx and Engels, according to which there are the contradictions originated by productive forces or by human labor that might account for contradictions existing in the spirit or human reason. We are going to see later on that we can introduce the Darwinian concept of evolution of the species, including man, as a valid alternative to the Hegelian viewpoint, or to the anti-Hegelian Marxism of the orthodox Marxists.*

From the point of view of a scientist trying to understand the intimate mechanism of human evolution in an accessible language, let us see how far an idealist conception as Hegel's can take us and still be a valid interpretation of human evolution in universal history. If the decisions that lead to the great transformations of history are taken by the leaders of the moment † we may think that such a decision results vectorially of the sum of all decisions of individual consciences, and if such a vectorial sum exceeds all other less popular decisions, the dominating power of historical transformation depends upon the decision of the great leader of the moment, i.e., of the sum of wishes and anxieties of the individual consciences.** (*See* Note II.)

Therefore, Hegel could be correct when he postulated that the evolution of history is a result of the manifestations of the general contradictions generated in the conscience of the people (*Volksgeist*). We might even build up a mechanical model in which Hegel's idealism might be transformed into a most trivial historical materialism. This may not have been far from Hegel's own thinking when he devised a system in which the universal conscience was his most concrete conception concerning the human spirit (or mind). As a corollary we might even deduce all other consequences that the "finite is the abstract," and the "individual conscience is the essence of idealism," in opposition to the "realism and concretism of the universal conscience" calculated as the vectorial sum of all individual consciences. Our interpretation, although somewhat simplistic, might constitute a better bridge to historical materialism than the "anti-Hegel" inversion postulated by Marx in his *German Ideology* (*see* Note I), and in all of Marx' and Engels' later manuscripts in which the universal conscience became the object and not the cause of transformations resulting from the contradictions in socioeconomic factors throughout history.

Therefore, the interpretation given above making Hegelian dialectics more palatable without the invertion of the formula into the anti-Hegel of the orthodox Marxism, might allow the establishment of a direct relation-

---

*This is despite the common myth that Darwinism may be applied to natural phenomena as studied in biology, and that historical materialism is peculiar to the evolution of human beings in a society ruled by sociological principles.

† Hegel used to identify the great leaders as the reason itself, and in a certain passage of his writings he refers to Napoleon as the "Reason riding his white horse"!

**This definition of leader may exclude most of Latin-American dictators, but would not exclude Hitler or Franco and not even Pinochet.

ship between the concepts of Hegel's philosophy, or the phenomenology of the spirit, and those that characterize historical materialism that became the basis of any violent or pacific reforms or alterations of the modern bourgeois state.

Furthermore, we know that Engels in his famous *Dialectic of Nature* (Note III) practically ruined his philosophy, or at least rendered it unpopular among scientists, by trying to extend the philosophical basis of Hegelian dialectics to the natural phenomena, from which great divergences arose between the followers of historical materialism and the natural scientists. The situation reached a dramatic climax when a whole crop of wheat was ruined in the Soviet Union by the inability of it (the wheat) to obey the postulates of the Marxist or Hegelian dialectics. (*See* Note IV.)

The reason is simple and could be explained by the model presented above of the vectorial forces that command evolution in universal history. And here we hit a neuralgic point. If history as a product of human activity could be reproduced by a vectorial model as the sum of the states of consciousness of the individuals of a certain population, science or scientific truth in no way could be demonstrated or understood by the sum of opinions of scientists however competent they might be. If we apply the mechanical (vector) model that we have suggested for history that could account for many situations (French Revolution, the rise of the bourgeoisie, Russian Revolution, the rise of Nazism, and its destruction by the Allied Forces) in no way could such a model be applied to the evolution of scientific thought or knowledge.

If Copernicus* initiated the scientific revolution with his model of the universe, this was achieved against the generalized opinion that the earth was the center of the Universe, and that the sun, the planets and stars revolve around it. It mattered little that somebody else could have agreed with Copernicus because the vector representing the ideas of the large majority was directed according to the opinion stressed by the Church that the earth was the center of the Universe. Finally, the work of a few (Gallileo, Newton, Laplace) convinced the majority of the learned men and erudites of the time that Copernicus was on the side of truth.

We know that Bacon himself, one of the founders of modern scientific attitude, cast doubt on the veracity of the Copernician model, on the basis of the simple fact of observation that if our planet turned around, all objects on it would fly in the opposite direction. Today any high school student can explain why such a situation does not happen, not because the majority of the spirits had learned such a trivial truth, but because a man of genius, Gallileo in the seventeenth century, had imaged a conclusive,

―――――――――
*T.S. Kuhn, *The Copernican Revolution*, Harvard Univ. Press, Cambridge, Mass., 1957.

although mental, experiment which could prove his principle of mechanical relativity. (*See* Note V.)

Let us get deeper into the great rivalry between abstract and concrete, using the example we were analysing of the apparent opposition (many think of identity) between what Marx and Engels (1857) called historical materialism which tried to subordinate history to the interplay of the forces of production and division of labor, and what characterizes the nebulous and difficult philosophy of Hegel, in his phenomenology of the spirit and his dialectical logic establishing that history and human evolution are the result of forces immanent to the human conscience itself, the contradictions of which would explain the great struggles of universal history. As we have seen, Hegelian dialectics may be adequate to explain social and human phenomena, since the instrument, the motor, behind such phenomena results from the algebraic (or vectorial) sum of all tendencies existing in the human units that constitute a community in evolution. In such a case it is understandable that the "reason" of a social phenomenon may result of all tendencies of the individual consciences.

In his poetical though nebulous style, Hegel was consistent with his conceptions of the *phenomenology of the spirit* and *dialectical logic* about human understanding and all categories the existence of which have been postulated as immanent to the human mind. When he refers to the equestrian figure of Napoleon Bonaparte, as "Reason riding his white horse" he was summarizing his basic idea that the forces directing the evolution of mankind were subordinated to reason that would bring in its context the dialectical contradictions that manifest themselves in the great strifes and wishes of history. To Hegel the universal conscience bringing with it all its affirmations and negations or internal (dialectical) contradictions, would plunge into history which would acquire all configurations (reliefs and counter-reliefs) of this universal conscience commanded by reason. The latter would materialize in the personalities of the significant leaders of history, this justifying his allegory of Napoleon as "Reason riding his white horse."

All different historical stages, including slavery, tyranny, democracy, freedom and dependency of some in relation to others, would constitute other manifestations of the conscience as a factor in the evolution of mankind, from the time men have abandoned their animalism to become citizens of large communities under the direction of significant leaders as the real personification of reason that dominated ever since the course of universal history. Some apparently paradoxical notions as the identity of the "being and the naught" *(l'etre et le néant)* of the "I and the non-I" did find their philosophical basis in the dialectical logic which acquired its

---

*Garaudy, 1966.

modern Marxist formulation in "thesis-antithesis-synthesis." As the "antithesis" is the negation of the "thesis" the "synthesis" would be the "negation of the negation," and so would be explained the great conflicts of conscience and by projection, the great conflicts of the history of mankind.*

It was not, of course, the first time that this strange concept of the identity of the "being and the non-being" was formulated in opposition to the empty concept of what could be called a tautology of $A=A$ or the iterative one, $A=B$, $B=C$, therefore $A=C$ with the exclusion of the paradoxical one A = non-A. In reality since Heraclitus †, in the fifth century B.C. the philosophy of the A = non-A had been formulated in face of the concrete fact that "the river in which I bathe myself today is no longer the same one in which I bathed myself yesterday," though its appearance be the same and its existence as the same river could not be put into doubt, not even by those endowed with the acutest power of observation; even the philosophical question of what remains of the egg transformed into the newborn chick which could not contain anything more than what can be found in the egg, but in which we could no longer distinguish the yolk, the white, the generative spot, that all together as an harmonious wholeness has been transformed in the chicken *in status nascendi*. To call the egg the "thesis," the embryo, the "antithesis," and the chick the "synthesis" would constitute a clear cut example of Hegelian dialectics. When we think of the transformation undergone by the feudal monarchic state in the democratic bourgeois state by virtue of the French Revolution, we are still contemplating the transformation of the "thesis" into the "antithesis" and finally into the "synthesis" implanted by Napoleon who represented the triumph of reason in the form of the constitution and laws of the bourgeois state (*codice napoleonici*).

This way of explaining evolution can only be a general idea, and certainly could not be considered a scientific method for the verification of the truth of concrete phenomena (natural phenomena). It is a truism that may have its equivalent in the idea that the form is always bound to matter, despite the fact that form has nothing to do with matter; or then that the spirit (soul) cannot be dissociated from matter (the brain) because when the latter is undone or desintegrated, after death, the spirit is likewise undone, unless we adopt the attitude of accepting, as many do, the survival (permanence) of the soul after death.

*H. Lefebvre, *Pour connaître la pensée de Karl Marx*, Bordas, France, 1966.

†To understand well the origins of Hegel's ideas nothing is more instructive than reading his monumental book: (G.W.F. Hegel), *Leçons sur l'histoire de la philosophie: La philosophie Grecque*, 3 volumes, published in 1833 and reedited in 1971, by the Librairie Philosophique, J. Vrin, Paris, under the auspices of CNRS of France. On p. 154, Hegel says: "There is no single proposition by Heraclitus that I did not retake in my logic."

Nonetheless, something tells us that the dialectical contradiction of the A = non A, i.e., that each individual is the negation of itself, and the synthesis "the negation of the negation" contains something more explicative than the mere tautology that "the form depends upon matter" or that "there is no matter without form," or that $A = A$ and cannot be its contrary, the non-A (as in the philosophy of Parmenides). This something more, existing in the dialectic logic, was largely exploited by the followers of historical materialism, and became the phantom menacing the bourgeois society frightened by the idea of transforming itself in its opposite, the non-bourgeois or proletarian state.

We have not to believe in everything that was prophesized by Marx and Engels in the second half of the nineteenth century, but we can assume that the inversion of Hegel's formula of contradictions of conscience to be explained by the competition of material factors of production and division of work, has been a profound idea that attracted the attention of a large number of economists and philosophers after its formulation in the mid-nineteenth century. Its power of conviction can still be evaluated by all measures of security (and repression) adopted by the capitalist states to avoid its materialization in the so-called class struggle (*lucha de classe*).

The affiliation of Marxism to the philosophical ideas of Hegel can be a mere expositive resource because what Marx and Engels did was to apply Hegel's concepts to the concrete phenomenon of class struggle. But such an application to a concrete event does not exhaust all the possibilities or implications of the dialectic conception of Hegel's doctrines because even in a classless society that continues to progress, the Hegelian or Heraclian concepts of the identity of the "being" with the "non-being" would continue to operate, unless such a classless society desired by the orthodox Marxists nothing else could be than a hive of honey bees in which progress becomes unnecessary by the perfect adjustment of all casts (individuals) working under an egalitarian system to the common life in the hive.

If one thinks that the elimination of the bourgeoisie (capitalism) and the rise of the worker class as a unique class would solve all the dialectical contradictions, one would simply (undoubtedly) be reducing the Hegelian conception of the "being=non-being" into an inept tautology very different from the original idea of Hegel's philosophy or of the Heraclian concept of the evolutionary process depending upon the dialectical contradictions; that is, of the "negation of the negation" as an infinite or indefinite process of transformations leading to evolution (or progress).

Therefore, even in a classless society the dialectical contradictions continue to exist because they by no means limit themselves to rivalries between master and slaves, employers and employees, rich people and poor people, but would still exist inside of each individual even if their duties and rights in relation to the community become equal, because each one in its dialectical, internal contradiction can always find that their rights are more important than their duties, and the fight will continue inside of the only class constituting the new society. There always will be the beautiful and the ugly, the honest and dishonest, the good and the bad, the rich and the less rich, and so on, and each one of these dualities will generate the dialectical contradictions triggering all over again an intra-class strife for survival.

In another place (M. Rocha e Silva, *A Evolução do Pensamento Científico*) I did suggest that Marxism or historical materialism as factor of social evolution could be more genuinely affiliated to the Darwinian idea of the "struggle for survival of the fittest," than to the nebulous philosophy of Hegel. The advantage of this change of coordinates would be to extend the historical materialism beyond the limits of mankind, introducing the idea of the survival of the fittest (stronger) as a factor of evolution in the zoological scale, from the amoeba to man, and from the protohominids that must have appeared in the surface of the earth some millions of years ago (the latest estimates put to four million years or more the time of the advent of the first forms similar to man) till the man of the atomic or spatial age.*

But some (and many did) may argue about what would be the advantage of this change of coordinates, when it would be perfectly conceivable to apply Hegelian dialectics to the evolution of the amoeba to sponges, from sponges to annelids, from annelids to mollusks, to fish, and so on, to man. On each step of evolution we could establish the fundamental dialectical logic equation, assuming that the sponge is the "non-amoeba," the annelid the "non-sponge," and the man the "non-ape," and would still obey the fundamental postulate of Hegelian dialectics. We know that that was some of what Engels tried to do extending the dialectic logic to the explanation of all natural phenomena, but in a certain way in so doing we are getting farther and farther away from Hegelian dialectics, unless we assume that the amoeba is endowed with a conscience that mirrors the fundamental condition of considering itself a non-amoeba that would be the force for its transformation into a sponge or an annelid, and so forth. I would not swear that such an extravaganza would be forbidden to the philosophers because they feel free to extend their conceptions to the limits of the possible and rational, and still find followers to develop their ideas. But what we really have done by applying Hegelian dialectics to the evolution of living bodies is a kind of description of the natural phenomena but not properly an explanation of the same.

*For a complete discussion of the origins of man, see Richard Leakey, *Origins*, 1977.

CHAPTER II

What has been introduced by Darwin's conception of the "supremacy of the fittest" is just the power force that can be identified in the living world from the amoeba to man in the fight for survival of the fittest inside the biological or ecological conditions of existence. Though Hegelian dialectics assumes the possibility of a struggle for survival, Darwin's theory gives the instrument (tool) acting to allow evolution to be achieved. This is the difference between saying that the individual (man) nourishes itself (himself) because it is hungry, from the explanation that to be properly used the food must undergo the action of all biochemical and physiological machinery of enzymes and movements of the digestive tract.

With such a confrontation of the concrete and the abstract, starting from the philosophical discussions of the empiricisits and mentalists of the seventeenth and eighteenth centuries culminating with the attempt (apparently successful) to adapt Hegel's idealism to the historical materialism of Marx and Engels, we were naturally led to a problem of the highest scientific importance, namely that of the evolution of the species, including man in his origins from the primates.

We arrive then to one of the crossroads of the history of scientific knowledge, in which we have to decide between a general philosophical survey (concept) as the one of Hegel's dialectic which generated its contrary, historical materialism, and a conception that we might denote as concretist, of analysing objective data of observation of the living bodies, such as all data accumulated by Darwin in his trip abroad the *Beagle,* and also data which were accumulated in the field of comparative anatomy (Cuvier) and in geology (Lyell) which constituted the raw material for Darwin's theory.*

We could image what would have happened if Darwin, instead of his conception of the "struggle for life and survival of the fittest," had presented his theory of the evolution of the species, including the evolution of primates into man (hominids) in the Heraclian or Hegelian philosophical language of the identity of the "being and non-being" or even in the more modern form of the thesis-antithesis-synthesis as sketched above. Only few biologists (if any) would have given the slightest credit to the best-seller of scientific literature of the nineteenth century, whose first edition was sold out in a few hours: "*On the origin of the species*" by Charles Darwin (1859). (*See* Note IV.)

What characterizes a scientific theory is its great capacity to suggest or stimulate new ideas on the basis of new observations or experiments. With such successive additions, the scientific theory transforms itself, improves or is eliminated in the course of time. In that dwells the great difference between a scientific theory living of the glory of being renewed, and a philosophical system which by its proper nature is immutable or eternal.

---

*See Darwin, F., 1892. Republished in 1958, by Dover Publ. Inc., New York.

In that, Darwin's theory does not differ from any other of the great theories of biology or of any natural science. With the discovery of the laws of transmission of the hereditary characters, since Mendel, De Vries and Morgan, it became apparent the insufficiency of the Darwinian conception of evolution by the simple struggle for life. It became necessary to introduce another factor, the "sudden mutation," to escape the deadlock that the simple usage or elimination of hereditary characters could bring about in the transformation of the amoeba to man. No doubt natural selection still is the most powerful factor for the fixation of the fittest to its ecological nest, but simple selection that gave so good results in the breeding of a pure-blood English pony, or to new races of corn and wheat, only exploits the already existing hereditary patrimony, when the great evolutionary factor, the transformation of genes by sudden mutation, does not occur.

The large number of observations about mutations of the fruit fly (*drosophils*) and of cultivated plants, brought about the improvement of Darwin's theories, and gave origin to such a monument of modern biology, namely the deciphering of the genetic code (*See* Beadle and Beadle (1967) as well as the last chapter.)

Obviously, whatever the transformation undergone by a scientific theory, it will be in agreement with Hegelian dialectics. This was the reason that led Engels to assume that all natural phenomenon obey their evolutionary principles. But it is just such an unlimited application of Hegelian dialectics and of its counterpart, the historical materialism, that weaken their power as a scientific explanation, which resulted in the discrediting of Engels concepts, as put forth in his *Dialectics of Nature*, among natural scientists.

As a first conclusion in this search for the origins of the scientific thinking we could simply exclude Hegelian dialectics and *a fortiori* historical materialism as basis for the scientific method, with the generality suggested by Engels in his *Dialectics of Nature*. Though trying to escape the tautology $A = A$, the Hegelian dialectics, by its generality falls back into another tautology when applied to the genesis of the scientific theory.

What is important for the scientist is not to know that the chick is the negation of the egg, but essentially to know the sequence of reactions that lead to the transformation of the egg substance into the substance of the chick.

Another great danger of Engel's conception is to submit the scientific knowledge to the existence of political ideologies. Such a subordination has been as repulsive in Nazi Germany where Chancellor Dolfuss was eliminated for presenting all the symptoms of "hypophysary cretinism," as it was nefarious to apply the Marxist-Leninist dialectics to justify the badly controlled experiments by Lysienko about the genetics of wheat and domestic animals. In one or the other case, the misunderstanding of scientific ethics (honesty) by the Nazis, and the absence of proper scientific

CHAPTER II 41

controls in the experiments of geneticists of the era of Stalin had ominous consequences: the hecatomb of the Second World War in the first case, and the ruin of the wheat harvest in Soviet Union in the second case.

Note I

The inversion by Marx and Engels of Hegel's Dialectics

"Though the French and the English at least adhere to the political illusion which is the nearest to reality, the Germans prefer to move in the domain of the 'pure spirit' (reason) and consider the religious illusion the motor of history. The philosophy of history by Hegel is the last consequent expression brought to its 'purest expression' (form) of the way by which the Germans write history, in which real interests are not concerned, not even political interests, but only 'pure ideas'... It is not the critics, but the revolution itself that constitutes the motor power of history, religion, philosophy, and of any other form of theory. This conception shows that the objective of history is not that of resolving itself in a 'conscience of itself' as the 'spirit of the spirit,' but in each stage one must consider a material result, a sum of the productive forces, a relation with nature, and between individuals, historically created and transmitted to each generation by the antecedent one, a mass of forces of production of capitals and circumstances which are in part modified by the new generation, but in part are dictated by the norms of existence themselves... consequently, the circumstances are created by men, as much as men are created by the circumstances; such a sum of forces of production, of capitals, of forms of social relationships that each individual, and each generation already find as pre-existing data, is the concrete basis of what the philosophers called the 'substance' and the 'essence' of man..."
Marx and Engels, *L'Ideologie Allemande,* 1846, reedit. by Editions Sociales, Paris, 1972, pp. 81, 79.

Note II

Hegel's Idea of the Volksgeist

"We have the tendency to call *people* the aggregate of private persons. But such an aggregate is the *vulgus* not the *populus,* and under this aspect the objective of the State consists in not allowing the people to exist or exercise its power under the form of an aggregate. A people that would be found in this situation would be a people in frenzy (delirious), a people in which would dominate the immorality, the injustice, the brute and blind forces. Would be like an unchained (*déchainé*)

ocean... The reason for the participation of the private individual in public affairs must be found, in part in his most concrete and intimate feelings of the most generalized necessities, but especially and essentially in the rights of the general spirit of the people *(Volksgeist)* to manifest externally, under the form equally general by a determined and well oriented activity, his will (desires) to participate in public affairs... It is not in the inorganic form of the individual as such by the democratic process of choice—but as an organic force, as member of the State that the individual should intervene in the public affairs.
G.W.E. Hegel, *Morceaux choisis,* Gallimard Edit, Paris, vol. 2, p. 230.

Note III

F. Engels, *Dialectic of Nature,* Progress Publishing, Moscow, 1934.

Note IV

T. Lysienko, *The Science of Biology.* International Publishers, New York, 1948. The book of the academic Lysienko that promoted such a scandal among occidental geneticists (scientific world) embodies his famous speech given in July 3, 1948, as a president of the Agricultural Academy of Sciences of the USSR. The position taken by the academic Lysienko is one of frank opposition to the basic concepts assumed by the majority of the geneticists both inside and outside of the Soviet Union, and also by the majority of the geneticists of Soviet Union, many of whom were fired from their positions by virtue of the political prestige assumed by Lysienko in the entourage of Stalin. The book was reviewed in volume 4 of *Ciencia e Cultura* (1949), p. 226, and commented upon in the editorial of the same number of the periodical, under the title "Science and Politics," p. 163.
See also Medvedev (1971).

Note V

The Principle of Mechanical Relativity

We refer here to the mental experiment suggested by Gallileo to let a weight fall from the top of the mast of a ship in motion: its trajectory would not be affected by the rectilineous movement of the ship. This experiment was apparently done after Gallileo's death by the philosopher Gassendi, and served as the basis to introduce the so-called Gallilean transformation that remains invariant in the case of the linear and uniform movement: "All mechanical phenomena (acceleration, force, impulse or moment) are independent of the rectilinear and uniform motion of coordinate axis to which they are referred to." This invariance led Newton to formulate his laws of universal gravitation. Only in the beginning of the twentieth century was it demonstrated that electrodynamic phenomena do not obey Gallileo's postulate with the advent of the relativity theory that revolutionized modern physics, based on the proposed Lorentz' transformation as an interpretation of the negative experiment by Michelson and Morley (1887). The latter was an attempt to apply Gallileo's transformations to a beam of light emitted by a fast moving body, such as the earth surface in its diurnal movement (if this experiment had been performed in the seventeenth century, probably Copernicus and Gallileo would have been defeated and the Church considered right).

Note VI

Scientific Method and Dialectic Logic

The comparison between the "method" of a naturalist, such as Darwin, and all those who have accumulated data on comparative anatomy (Cuvier) or on the geological transformations of the earth crust (Lyell), that served to substantiate the *hypothesis* of natural evolution of the species, including man, pronounced almost simultaneously by Wallace and Darwin, and the Hegelian "method" of conceiving universal history as resulting of contradictions of the human mind, would constitute an excellent subsidy to understand the eternal opposition between concrete and abstract, empiricism and idealism. In the case of the evolution of the species from the amoeba to man, the balance would certainly incline in favor of the empiricism or Darwinian concretism, especially as a consequence of the tremendous success* of the book *On the Origin of the Species* by Charles Darwin (1859). In the case of the human evolution along universal history, it was necessary to invert Hegelian dialectics, extricating from it all that was purely idealist, introducing the idea of "class struggle" and the contradictions of the forces of production, into what became celebrated as historical materialism, Marxism-Leninism, Maoism, that inflamed a great part of the population of the globe. It appears, however, evident that Marxism and Darwinism pursue parallel courses assuming as the motor force of the progress the characteristics of human labor, i.e., its capacity to manipulate the means of production, invent and utilize instruments, activities that are barren to the animals. But the concept of the survival of the fittest and sudden mutations only tangentially appear in the basic philosophy of historical materialism.

*The first edition was published in Nov. 24, 1959, and the 1250 copies were sold in the same day. See F. Darwin (1892).

Chapter III

MYTH AND SCIENCE: THE ORIGINS OF SCIENTIFIC KNOWLEDGE

III
MYTH AND SCIENCE: THE ORIGINS OF SCIENTIFIC KNOWLEDGE

When we depart from Hegelian dialectics as a scientific method, we open a kind of vacuum in what may be called the origins of scientific thinking. If we exclude reason as the only source of knowledge, we also exclude that voracious conscience postulated by Hegel as the force that penetrates and encompasses the world of external events: the objective and the concrete. To assume that there are these contradictions of the mind (spirit) or universal consciousness that constitute the motor behind the evolution of science is another way (tautologic) to say that the evolution takes place and that everything evolves because it contains in its context the "non-thing" that finally transforms itself in the final product that would be the negation of the "non-thing" and so on. As we have seen, such a way to interpret evolution may have its place when dealing with universal history, the makers of which being the great leaders who interpret the dominating tendencies in the society (or community) where they emerged.*

The transformation of the monarchic and feudal society into the bourgeois society that arose from the dialectic contradiction between the classes of the French society of the eighteenth century found in Napoleon its most powerful instrument or its "efficient cause." This justified the enthusiasm of the young Hegel who interpreted such events as a crystalline example of his ideas of the existing contradictions in human conscience. Napoleon became the personification of the "Reason" itself, as we have seen in the previous chapter. It was this triumph of reason that permitted the implantation in all Europe of a form of enlightened and constitutional

*H. Lefebvre, *Morceaux Choisis de Hegel*, Gallimard, Paris, 1938.

monarchy, the model of which, though imperfect, would be the Prussian state, to which Hegel himself sketched a constitution according to his *phenomenology of the spirit* and *dialectical logic* (*See* Note II).

But the same argument could be applied to Nazi Germany that found in Hitler its instrument or "efficient cause" for the realization of the Third Reich, or else, in the Allied side, the three leaders Roosevelt, Churchill and Stalin, not to mention De Gaulle, as the "efficient cause" of the destruction of the Nazi empire that should have lasted a millenium, and crumbled in a little more than 15 years. Therefore, here also the Hegelian dialectics would not allow a forecast of the end results, but simply to depict *a posteriori,* the course of the events.

Even if we adopt the inversion of Hegelian dialectics into the historical materialism, it is doubtful that the fundamental contradiction postulated by Marx and Engels between workers and employers that would lead to the "class struggle" and the setting up of the classless society, have really been the decisive factor in the emergence of the socialist states that spread out today over one third of the civilized world.

One could hardly assume that it was such a fundamental contradiction between proletarians and capitalists what lead to the victory of the Russian Revolution, or that of Mao-Tse-Tung, or of Ho-Chi-Min, or of Fidel Castro, to give the most genuine examples of the transformation of the bourgeois state into a basically proletarian one. We will not discuss here if in all such cases of victory of the Marxist cause the transformations have resulted from contradictions of the material forces of production, in accordance with the postulates of historical materialism. What we may discuss, and even doubt, is whether such transformations resulted from a genuine class struggle, as that prophesized by the orthodox Marxists. In an earlier work, we have discussed in more detail such particular case.*

If we pretend to apply Hegelian or Marxist dialectics to the evolution of science, then we encounter a much higher barrier than that which we found to explain social evolution. We certainly can analyse what can be concrete or abstract in the evolution of scientific knowledge, and here we can visualize the forces that press science forward, in what is most genuinely spiritual, the capacity of the human mind (intellect) to generate a scientific theory.

In another place (**) we tried to establish a parallel between invention in science, and the creative activity in plastic arts and literature, and have reached a small impasse. Creativity that is free in most creative arts, in music, in literature, must be submitted to a limitation that refers to the *scientific method*. In that sense, scientific creativity cannot be taken as a

*M. Rocha e Silva, 1976. Conference in UDUAL, Mexico, Loc. cit.
**M. Rocha e Silva (1971).

scientific method, since it is free to operate up to the moment in which it must submit itself to a verification of the veracity of the dreams (or fantasies) of the scientist that developed the new theory, and at this point enters the scientific method which cannot be ignored by any scientist who pretends to further his science.

But let us see how the duality, abstract-concrete, presents itself in face of the scientific knowledge.

The scientist ignores the abstract for a question of principle, of commodity, of education, or simply of tradition. The abstract represents the abyss in which he can plunge without any hope of salvation, though the concrete constitutes the firm ground where he can put his feet, to feel by contact the fluidity or solidity of the ground in which he steps, then his tendency to pay attention almost exclusively to the concrete.

However, in the world in which he lives, the most powerful instrument to establish communication, social relations or to describe the universe is the language that contains a substantial part of the abstract.

Not seldom, the individual feels the difficulties to decide if what he sees, touches or hears, is abstract or concrete, for instance, in face of a mirage that suggests something in the sands of the desert, and becomes blurred and disappears when one gets nearer and nearer. The common language contains expressions that describe such a situation: "to take the clouds for Juno," "the saint for the sinner," or many other expressions that show that even the most concrete things can be erroneously interpredted by the touch, the taste, the sight and the hearing. From another point of view, if one analyses the concrete that presents itself to the scientist, the errors committed in their interpretation are of such a stature to put in doubt the infallibility of the concrete in relation to the abstract.

Let us consider first the abstract in the common language and in the language of the scientist. The world is interpreted *in abstracto*. Even the most concrete human actions may receive an interpretation *in abstracto,* that may be part of a theory about their motives, objects and consequences. Despite all the recommendations of the behaviorists who try to analyse as a *fait accompli,* a fact in itself, that must be only considered in relation to the probability of occurrence or of "emission of behavior," the tendency is to resist such a rigid (concrete) interpretation and look for some "abstract" motives that may explain it.

The psychoanalysts who were the first behaviorists of the end of the last and beginning of the present century are shocked by the attitude of the new behaviorists of the Skinnerian school to repel the abstract, subconscious motivations (as a taboo) to the experimental behaviorists.

To the abstractionists, animals, including men, drink because they are thirsty, eat because they are hungry, though to the behaviorist (Skinnerian) the animal as well as man have thirst because they drink, and hunger because they look for food; and thirst, starvation, deprivation of the sexual contact could be only evaluated by the high or low "probability of emission of behavior" searching for water, food, the female or the male.

It was such a repulsive feeling about the abstraction of the theoretical interpretations that led Skinner to reject the concept of the *Homo autonomus* and try to develop a technology of behavior, in which the latter and only the latter could be the object of scientific enquiry. With that, with such a concretist attitude we may even think of a human society as utopia in which all human attitudes would be based on the idea of "probability of emission" of a certain behavior.*

It was not without a certain degree of justice that the Skinnerian utopia was considered fascist or at least totalitarian by one of his most distinguished opponents, the mathematician linguist N. Chomsky. †To the dictators, what is important is the behavior (or better, the good behavior) of their subjects, and to obtain it they also apply a technology of behavior that consists, as in the case of Skinnerian technology, to submit their subjects to a process of learning, in which the "reinforcement," be it reward or punition, is utilized to obtain the "optimum emission" of the desired behavior.

But, it is not this kind of criticism to the Skinnerian technology that will concern us at this moment. Our objective is more fundamental, that of trying to justify both attitudes (after all there are more fascists in the world than democrats). The simple fact to ignore them does not exclude the fundamental duality existing in both universes in which we live: the one of *appearance,* and the other of *reality.* Not infrequently, the two universes present themselves as complementary forms of the same thing, and it is the task of man (scientist, artist, or simply the common man) to fuse or bind with a dash-the two aspects of the world in which we live.

It has always been the dominating argument of philosophies of all times to try to establish the relationship that must exist between such worlds, and we find a long list of names and concepts that are usually presented in pairs, as translating into different contrasts that fundamental duality: "classic and romantic," "cerebral and emotive," "idealist and empiricist," "introverted and extroverted," "visionary and realist," "abstract and concrete," as dominating characteristics of the one and the other aspects between cultures, some already extinct, or between peoples or races still existing or on the way toward extinction.

The individuals themselves may present marked characteristics that classify them between conservatives and subversives, idealists or practical men, romantics or classicals, reactionaries and progressive, extroverts and

---
*B.F. Skinner, *Walden Two,* 1968.
†N. Chomsky, *For Reasons of State*, 1974. Loc. Cit. Chapter 1.

introverts, and so forth. Many tried to establish relationships between somatic parameters and psychological qualities, being the fat, brevilineous, brachycephalous, in general, of the concrete, conservative, extroverted (easy communication) type, though the lean, longilineous, dolichocephalous, would predominantly be of the romantic, idealist, introverted, and abstract type!

"Let me have men about me that are fat...
Yond Cassius has a lean and hungry look;
He thinks too much: such men are dangerous." *

All those who are familiar with the classification by Kretschemer (*Physique and Character*, 1952) would also be reminded of the athletic type, with large shoulders, a sculptured face, presenting tendencies of the intermediary type in which the concrete combines itself with the abstract, generating a temperament sometimes aggressive, sometimes sweet, the maniac-depressive type, in which the crises of depression alternate with the phases of extroversion all over its life. There are always the classical examples: Goethe (extroverted), Schiller (introverted), Nietzche (athletic), and each one can spot in their own environment representative types of Kretschemer's classification.

We may assume a great deal of arbitrariness in such classifications which have rather a clinical significance for the diagnosis of certain alterations of the personality, but the objective we have in mind is not properly that of trying to understand or justify the attitude of the psychiatrist when he proposes his diagnosis, and even less to sketch a therapeutic scheme for such deviations of the personality. Our objective is to search, in language and science, for traces of the psychological dualism that probably exists in each of us, and is a result of the position of man placed between two universes that oppose themselves, but are at the same time complementary.

The objective that we are trying to attain is that of encountering what is concrete and abstract in human science, starting from the possible analogy that may exist between the dual mechanism by which the individual acquires the capacity to make use of the mother language, and his apprenticeship of science. But, more than that, to establish relationships between what the linguists call the generative grammar or transformational grammar (TG), and scientific creativity.

We know today that the function of a grammar is that of proposing the fundamental laws (or rules) that command the use (performance) of a language, but the great contribution by Chomsky and his school,† has been that of showing that with a limited (or finite) number of grammatical rules, the individual who dominates its mother language is able to create phrases

*W. Shakespeare, *Julius Caesar*, Act I, Sc. 2.
† N. Chomsky, 1966, 71.

or ways of expressing himself that by no means are explicit in the grammar of his (mother) language. In other ways, with a finite number of laws or rules, the individual is able to generate an infinite, or at least indefinite number of phrases, among which some may have been uttered for the first time, without any specific learning process being needed to the formation of the new phrase. *(See* Note I.)

Here we encounter the great antagonism between Skinnerians and Chomskysts. For the former, the individual may be confined to a learning process (involving only external contingencies) in which a mechanism of reinforcement is necessarily involved, and the acquisition of knowledge (performance) can be measured by the "probability of emission" of a certain "verbal behavior."* The knowledge of the grammar would come *a posteriori and would serve to rationalize the acquisition of that "verbal behaviour." For Chomsky and his more faithful followers, grammar would be something more profound and inherent to the genotype of the individual. It would be a pattern* pre-existing in the human mind, somehow as the archetypes of Jung, provided the latter are understood as a possibility, more than an existing real structure of the "collective unconscious." *(See* Chapter I, Note V.)

This point deserves an additional explanation. If in a dream the individual profoundly impressed by religious myths has a vision of St. George stabbing a dragon, this would not mean that this image created in the Middle Ages by the inventors of such religious myths has been transmitted as such, from generation to generation, through the *collective unconscious,* as naively believe some of the interpreters of Jung's ideas. Such an explanation could not find a basis in the modern theories of genetics. What the individual, as a representative of the species *Homo sapiens* could have inherited from Christian mythologies of the Middle Ages, would be solely the capacity to interpret, by means of more or less fantastic pictures, the anxieties and fears originated in the biological struggles for survival in a world that can be hostile, and would be part of the structure of the *Homo autonomus* so emphatically denied by Skinner and his followers, that could constitute the basis of Chomsky's ideas about the acquisition of one's mother language.

If the individual awakes terrified and says: "I dreamt that St. George rescued me with his sword from the clutches of a dragon!" he might be emitting this phrase for the first time, and the situation described might have been unique in his life, that of being menaced by a non-existing monster, and being rescued by a supernatural force, in that case, the image of a St. George armored from *pied en cap.* Notwithstanding being unique and not inherited from his ancestors under the form of an image pre-existing in his collective unconscious, the individual can emit, for the first

---

*B.F. Skinner, 1957.

time, a phrase that might describe with precision his fear in face of the world in which he lives. In the interpretation of his dream, the dragon may symbolize the hostile figure of his boss at the office, or of his wife, or of his mother-in-law in the family environment, or else the scene visualized in his dream, may represent a difficult situation in face of urgent obligations.*

The image itself, of the dragon stabbed by Saint George was introduced as a psychological predetermined *pattern,* that could be the same as that of the fathers of the Church in the Middle Ages, tormented by anxieties and desires which could not be satisfied in their restricted worlds at the convent cubicles. But the new phrase probably emitted by the first time and never learned before, has something of such linguistic patterns so often mentioned by Chomsky. The uttered sentence is only one of the infinite variety that could be emitted in such an opportunity, all of them translating into a mental pattern belonging rather to the species than to the individual, but that would exteriorize itself by an oneiric image and by a linguistic interpretation of an unexpected modality.

We do not want, by that, justify the mentalistic attitude of the Chomskysts, nor exclude the Skinnerian interpretation, but wish only to show that the duality that we have been stressing is part of the intimate life of anyone, and may appear in the descriptions that we make of the world of reality or of the world of fantasy of the dream. It is the approximation of these two worlds in a mentalistic plan, between dream and reality, that justifies to a certain extent the interpretations of the protagonists of the generative grammar (TG), of attributing to the latter the qualities of *insinuation* more than of teaching, the general orientation that must be given to the phrase (output) uttered by the speaker, and the competence (input) of the individual (hearer) to whom the speech is directed.

In a certain way, we could understand Jung's archetypes as translating in more or less fantastic images, the pictorial form of the universal language, as if the individual in his dreams could see a cinema screen that develops in space, something that can also be translated in a temporal sequence in his mother language as the "outline of the dream," or in more elaborate form, by the "interpretation of the dream".† We may also assume that this "interpretation of dreams," expurgated of all possible arbitrariness or extravagancies introduced by the psychoanalysts, and by all magicians who have interpreted the dreams since the most remote antiquity,† or still today, in the popular almanacs dealing with the "interpretation of dreams" can nonetheless contain something of strictly personal and "true" about the

---

*Incidentally, the situation could not be interpreted as resulting from a Skinnerian experience, by the simple fact that such a situation could never be observed, since dragons do not exist.

† S. Freud, *The Interpretation of Dreams,* 1972.

wishes and tendencies unsuspected by the conscience of the dreamer.*

Such an interpretation of dreams, as a coded (figurative) language, pictorial or symbolic, that can be *seen* and not really heard, and that presents the panoramic character of all manifestations of the unconscious, and that often presents the symbolism of the *myth,* as real sketches (comics), of the mythology created by the species, would have the same structure of the articulated language when expressed in pictograms or in the writings of the ancient Egyptians (hieroglyphics), Chinese and the Mayans. It is to be noted that myth in Greek would mean only "words" and mythology would be the description or the study of the "word," indicating that for the Greeks, the myths inherited from their ancestors could signify no more than words, as in articulate language: "Words, Words" in the Hamletian description of the content of books; or in the biblical connotation: "*primum erat verbus*," translated by Goethe as meaning: "in the beginning there was the action," (*Im Anfang war die Tat,* Faust, Act 1, Part 1).†

If we assume that many of the dreams with a collective meaning, in accordance with Jung's and his followers ideas (as presented in *Analytical Psychology)* develop themes that under many respects are mythological or can be interpreted as such (the dragon, the serpent, the hero, the monster, the sin, the sex, the incest, the god-child, the mother virgin) we can reach a first important generalization, namely that the mental organization of the unconscious (even of the child, translated into childs dreams), allows the articulation of symbols representative of a universal language obeying the rules of an unconscious grammar. And the efforts of the child to learn its mother language would be only that of translating into a phonetic (superficial structure) language (peculiar to each tongue) the infinite varieties of its unconscious constructs. (*See* Note II.)

If one accepts such a first generalization, the *mentalistic* concepts of Chomsky of the acquisition of the mother tongue loses a great deal of that character of arbitrariness that his critics (especially the behaviorists) attributed to him. Furthermore, if we may assume Chomsky's mentalistic attitude about the acquisition of the language as an acceptable possibility or perhaps a simple working hypothesis, we may find the paths open to understand this fabulous fusion of the abstract (substance of the dream) and of the concrete (events of everyday life), translated into a language so easily acquired by the individual since its earliest life. And we also open the pathways to understand the no less fabulous capacity of the human intellect to understand scientifically the world in which we live. (*See* Notes III and IV.)

*C. Jung, *Analytical Psychology*, loc. cit.

† J.W. Goethe, *Faust*, Act I, Part 1, Line 883.

In another work (M. Rocha e Silva, *A evolução do pensamento científico.*, Hucitec, 1972.) we tried to establish some analogy between myth and scientific theory, when it is being emitted or conceived for the first time, or in *status nascendi*. In order to accept such an analogy we should, on the one hand, valorize the myth as a genuine product of human creativity, and on the other, reduce to reasonable confinements the rationality, sometimes exaggerated, attributed to the creation of a new theory or conception about the intimate nature of the observed phenomena.*

In such an attempt to approximate the two concepts apparently so different, namely that of the myth and that of the scientific theory, there must enter a large dose of what the philosophers and metaphysicists call "idealization" or construction of "ideal types." These must adjust themselves to reality as much as the idea of a right triangle is adapted to the geometric picture sketched in the blackboard for the deduction in class of Pythagorium theorem. When the professor begins his deduction drawing lines that intercept themselves and affirm emphatically: "Let us consider a right triangle..." neither he himself nor his students are convinced that the three crooked lines that intercept themselves forming angles more or less obtuse or acute are the exact representation of a right triangle.

Here, we may recall the definition (one of the definitions) of geometry proposed by Poincare, as "the art to reason about badly drawn pictures."†
It would be, just to give another example, as the operation proposed by Max Weber (*See* Aron, 1967) to establish the relation between capitalism and the Protestant religion, establishing "ideal types" of individuals and communities that would present the essential characteristics of a capitalist structure, and the spiritual qualities of the Anglo-Saxon commonwealth which adopted the Protestant religion as an individual religion bound to the state. We know that such approximations are often very artificial, and are proposed only as "working hypotheses" for the scientific investigation, not only on natural sciences but also in sociology and theology. It is with such limitations that we may introduce such an approximation between myth and scientific theory.

Undoubtedly, the myth has something more than the arbitrary creation of stories and tales produced by the free poetic or literary imagination. In the first place, the myth reproduces itself under different forms in the most diverse cultures, in ways that we may assume to be independent.**The myth of the creation of the world appears in the Bible, in Egyptian hieroglyphics,

---

*M. Rocha e Silva, *A logica da invenção e outros ensaios*, 1965, *and also* A. Teixeira and M. Rocha e Silva, *Dialogo sobre a logica do conhecimento*, Edart Edit, 1966.

† H.Poincaré, *La science et l'hypothèse*, Flammarion, Edit., Paris, 1902.

** Levi-Strauss, 1970; also, E. Leach, 1974.

in the ideograms of the Chinese, the Mayans, and in the oral traditions of people from Africa, Asia, Oceania, and in tales by natives from Australia, and from North, Central and South America. The myths of the god-child, the virgin, the hero (who fights dragons, serpents, imaginary animals such as the unicorns, the centaurs, and so forth) went across millenia of verbal and written tradition, and still remain as the basis of religions of hyper-civilized people, in the Middle Ages, in the European Renaissance, or in the middle of the twentieth century, in Christianity under diverse patterns, in Islam, Buddhism, or in black magic. It does not matter that other myths of a scientific nature, as the evolution of the species, the atomic and nuclear theories of the constitution of matter, quantum mechanics, wave-corpuscles, transmission of heriditary characteristics, ribbons of DNA or RNA, are slowly supplanting the previous myths in the more cultivated minds of the man of the atomic age. Many of such better informed individuals keep in their mind many of such more primitive myths, as a kind of irreducible residue preserved in the innermost sites of their minds. And today through the activity of hippies and amateur exorcists, a wave of archaic mysticism appears to stir up the mentality of individuals placed in almost all walks of social hierarchy. (*See* Note I, Chapter I).

The coexistence of these two worlds, that of the religious myths, even in their most sophisticated forms as introduced by Christianity, Judaism, Islam, Buddhism or whatever name religions that are still followed by a large part of the population of the world might have, and the no less abstract or mythologic conceptions of quantum mechanics, force of gravity, four-dimensional space-time continuum, the genetic code, and so forth, shows the almost gigantic nature of the duality persisting in the human mind, of the two worlds we are speaking about, that of the fantasy and that of reality.

The attempt to exclude such a duality, by naive assertions, such that God is dead because the astronauts have never seen in their spatial trips anything that might suggest the existence of God, or on the other hand, the no less naive idea that God with his infinite knowledge planned all the intricate pathways of quantum mechanics when he created the atom, with all its elementary particles or the four-dimensional space-time according to relativistic relationships, or when creating the amoeba or the paramecia, already had a forecast of all possibilities of the combinations of molecules to the synthesis of proteins, or the possibility of transmitting information through the interaction of molecules of DNA or RNA, or everything that a legion of biochemists, physiologists, biophysicists, pharmacologists and pathologists, try to disentangle patiently to clarify the intricate nature of the living substance. All such simplifying or unacceptable assertions only proclaim that such a duality persists, and that something abstract must penetrate or try to explain what is of most concrete in everyday life.

CHAPTER III

One may give a few examples of how the living substance of the amphibians had to solve the problem of adapting themselves to life outside the pond water, or the reptiles solving their problem to acquire wings to conquer the aerial space, or of the primate to adapt itself to the erect life, to acquire the capacity of manipulating tools and finally to elaborate an articulate and symbolic language to describe the world in which he was living. Many will call this way of facing natural evolution as teleological, smelling of mysticism or metaphysics.

The danger of a teleological or metaphysical explanation resides not properly in its formulation, but solely in the undesired possibility of convincing mystics and adventurers that they may utilize it as the ultimate unconditionnally accepted truth.

The teleological explanation may have a sense when one looks through the window to admire the magnificent spectacle of nature in its accomplished form. We may object to the value of an interpretation that only admits as an instrument of work, the truism which states that nature or the natural phenomenon exists with such and such characteristics. The fact that appears obvious to us is that everything (all causes) converge to produce a certain effect, exactly the one that we observe in the real world. The hazards of such an attitude are that we may be forced to exclude the need for an investigation of the causes once we know the effects.

Denying any value of teleological explanations is part of the scientific methodology and must be enforced in any laboratory whose functions are that of investigating the causes of the natural phenomena.

However, for the common man, and even to the scientist himself when after finishing his work he goes out of his laboratory to relax, the explanation given to natural phenomena is almost always a teleological or metaphysical one, with the rare exceptions of the pedantic who insists on explaining details of a natural landscape, such as Niagara or Ignassú Falls, in terms of physical laws about the dynamics of fluids, or who explains the beauty of the rainbow on the basis of the laws of physical optics. (We should distinguish the attitude of a Newton discovering the causes of the rainbow, and that of the tourist or layman admiring the variety of colors in front of him, and so forth.)

Without going too deeply into that difficult subject, taboo to the scientist, we cannot escape the idea that natural evolution *did* actually happen and, therefore, had a cause unknown at one time, or *still* unknown and, such a cause was planned by the living matter, in a way similar to the Golden Gate or the Rio-Niteroi bridge, or the Metro of Paris (or the New York subways) or that of Saõ Paulo. We could, for example, assume that at a certain moment, amphibians developed lungs to adapt themselves to terrestrial life, reptiles developed wings to conquer aerial space, and the primates started to apprehend objects, handling them as tools, and developed an articulate language with all subtleties of the human language.

It must be assumed that at a certain moment that may have lasted several thousands or millions of years, such puzzles were solved by the living matter, either by an adaptive or "teleological" force, or by the blind play of molecular combinations that ended up in favorable mutations. However, we must assume that the puzzle or problem was deciphered by an organizational phenomenon, and that the "probability" of occurrence has been 100%, which excludes the simple participation of chance. (As when we look at the window and see that it is raining, and therefore the 100% rain forecast had been confirmed.)

In general, the teleological explanation is excluded with the argument that though the rare event such as the acquisition of a lung, wings or of the articulate spoken language has occurred, nobody would have forecast, and certainly nobody *did* forecast, such an occurrence, which weakens the notion that it was the pre-ordained act of some mysterious force of an omnipotent entity (though such a possibility cannot be excluded, it is incredibly improbable). Nonetheless nothing could prevent us from assuming that in the same way that a lung, a wing or an articulate language has been created by the act of creation during the evolution of the species, something that might have remained as a residue of it in the intimacy of the unconscious, under the form of a pictorial language, might give to the dreamer the elements to understand the external world, along the same line as the elaborated conscious language of the *Homo sapiens et loquaciens* is admirably adjusted to its function of describing the Universe, of drawing conclusions from well-formulated premises, in accordance with the computational logic of the consciousness.

The objection of the behaviorists and of all who avoid subjective or mentalistic explanations is obviously that of asking how could it be proved. Chomsky and his followers know that this primitive language could be used by the child to acquire the so-called linguistic competence, and this point has been focused upon by a number of recent papers and books, on one and the other side of the fence between mentalists and behaviorists. [*See* Chumsky (1966) and Skinner (1957).] It is quite clear that neither Professor Chomsky nor any of us would be in a position to decide that point, but it is nonetheless an interesting idea to postulate the existence of an universal form that might serve as a pattern to the learning of the mother language, as largely explored by Lenneberg (1967). (*See* also Langacker, 1972.)

The problem of course is not only linguistic. When the amphibian acquired better lungs to survive a terrestrial life, it was not a simple question of semantics that had to be solved. That would be the case in a science fiction novel, but the lung had to conform itself to certain structural specifications, those of a lung of a terrestrial animal, with trachea, alveoli, epithelia permeable to atmospheric gases, an appropriate respiratory musculature, an intrapleural void space, and everything else that allows a lung to function as a reptile, avian or mammalian lung. Such are technical problems of extreme complexity, that not even modern technology has solved in its entirety (properly), in a better way than by such uncomfortable and bulky iron lungs. The same could be said of the solution given by reptiles transforming themselves from crawling animals into flying machines, or birds. Only in the twentieth century did it become possible for men to solve the technical problem of flying heavier-than-air bodies, and the technological refinements used by the modern airplane factories still are far from the subtleties with which nature used to conquer the (aerial) space for insects and later to transform a heavy pterodactyl into a hummingbird or an albatross.

How the living substance was able to solve such problems we are only now beginning to visualize by the interaction of DNA and RNA in the synthesis of protein molecules. But the problem that is only glimpsed today by the scientist is only a shabby model of what the living substance achieved when it solved the practical problem, the puzzle of transforming aquatic animal into a terrestrial one, and a terrestrial into an aerial one, or of a primate unable to manipulate tools into the modern technologist who invents and manipulates the machinery needed to take man to the moon, and which has doubled human life span in the last two decades. What we may say to sympathize with the technologist is that they achieve now in a few years what the living substance took thousands or millions of years to develop under natural circumstances.

It is almost a truism to say today that the evolving tendencies intrinsic to the living matter has exteriorized in the *Homo technicus* under the form of scientific creativity, accelerating an evolution that took millions of years to reach pre-history, and a little more than 10 to 15 thousand years from that to the present day. The modern man completed such an evolution creating megalopolis, jet aircraft, radio, television, and antibiotics in the large conglomerates of the industrial and atomic age. The problems that are still pending, such as the poisoning of the air by all sorts of pollutions, should probably be solved extrinsically in a similar way to the problem of "interior pollution" that was solved by creating organs of excretion, the cloaca and the urinary apparatus, to avoid the retention of dejects of its own metabolism when it passed from aquatic to terrestrial life, or more precisely, when it became independent of the composition of the environment, creating its own "internal milieu," to defend itself against variations of salinity, pH and finally of temperature changes as homeothermic animals.

If the living substance was able to use millions of years to solve such problems of internal pollution, the modern man *must* solve his problem of external pollution in the next decades, or he will be exposed to universal destruction, as it would happen to the amphibians that would be obliged to live under conditions of terrestrial life with excretion organs adapted only to aquatic life. (*See* Note VI.)

Note I

That might be the main difference between animal and human language. The former consists of a limited number of sound signals (phonemes) with which the cat or the dog manifest their desires and anxieties. If the cat when fighting with a dog would use the language of the latter (*bow-wow* instead of *meow)* to manifest its hatred, it would be already in the way of the articulated human language, which with a much larger number of phonemes (from 20 to 45 in the occidental languages) is able to construct the infinite variety of expressions (or concepts) of the common human language. The other characteristic of human language is that of associating symbols to its capacity of forming new concepts, what allows it to distinguish again parrot's language from human language, despite the fact that many humans talk like parrots, and the parrots imitate the ability of humans to adapt their language to new situations that generate new concepts. What we cannot exclude from the animal world is the capacity of formulating concepts symbolized by attitudes and behavior patterns that we could call "Pauciphonetic" in comparison with human language. We may even classify as science fiction all the fables of Aesop or La Fontaine, as well as the comic sketches in which the animals can use a language similar to that of humans, with the introductory slogan: "If the animals spoke..." As has been said, under many respects the animals possess a wisdom superior to that of humans, as shown in their gregarious capacity and the absence of destructive instincts in relation to others of their kind, in opposition to the infinite stupidity of human beings, as exemplified by the last two world wars, and the wars of Vietnam and Middle East.

Note II

Some opinions about the nature of the myth

According to Schelling (cit. in Kerényi, 1971): "Mythology remains, for us, a phenomenon which in profundity, permanence, and universality is comparable only with Nature herself." According to Malinowski (1926, *Myth in Primitive Psychology*. Cit. in Kerényi, 1971) "The myth in a primitive society, i.e., in its original living form, is not a mere tale told, but a reality lived. It is not in the nature of an invention as read in our novels today, but living reality, believed to have occurred in primordial times and to be influencing ever afterwards the world and the destinies of man." For Kerényi (1971), mythology gives a ground, lays a foundation. It does not answer the question *why?* but *whence*. The myths form the foundation or ground of the world, since everything rests on them. They are the αρχαι *(archai)* to which everything individual and particular goes back, and out of which it is made, while they remain ageless, inexhaustible, invincible in timeless primordiality, in a past that proves imperishable because of its eternally repeated rebirths. The relations between historical events and mythological abstractions can be found in some well-known cases. According to some Etruscan inscriptions mentioned in Plutarch, Rome was founded starting from a circle (*mundus*) traced on the ground by the common agrarian instrument (a *plough*), and according to Ovid the first altar was situated in a *fossa*, the original site of the city of *Roma quadrata*. Now the combination of a circle with a square, that is a square inscribed in a circle, results in a mystical or cabalistic structure which appears in the Indian mythology and in Buddhism, with the connotation of a *mandala*. Buddhist temples are built-up in the form of a mandala, a square building with four lateral entries on the ground with the form of a circle, with the petals of a lotus flower irradiating to the four cardinal directions and the four intermediary ones. According to Jung, the form of a mandala would be an *archetype*; many individuals, without any knowledge of the history of religions, dream with the forms of a mandala, what in accordance with the concepts of Jung's "functional psychology" would constitute a psychic fact, autonomous and characterized by a *phenomenology* that repeats itself and is always identical when it appears.

See Jung and Kerényi (1971): *Essays on a Science of Mythology*, Princeton Univ. Press, Princeton, New Jersey.

Note III

We have gone from myth to reality of modern man, represented symbolically in his dreams, according to the psychological theory (analytic psychology) of Jung. Any other pathway would bring us to similar conclusions: that the anxieties, mysteries and myths that preoccupied the primitive man could still be present in the unconsciousness of modern man who can read by himself, or with the aid of a psychoanalyst, this primitive language in the form of "puzzles" (or enigmas) that the dreaming individual can frequently "read" in the context of its dreams some interpretations of the events of its everyday life.

If the code that must be deciphered by the dreamer may acquire some sense as a commentary of what happened in the days that preceded the dream, it may resemble a code or signal, with a connotation similar to an Egyptian hieroglyph or a pictorial symbol of the Chinese ideograms. Being a code, a signal, we may qualify it as belonging to a language, and being unconscious they could have the characteristics of the forms or moulds postulated by Chomsky and his followers, as a mentalistic substrate for the learning of any native language.

We could pursue such a line up to the point to provoke the revolt or objection of the behaviorists (Skinnerians), who see in the learning of a language nothing else than what is implicit in the learning of any operating behavior.

Nothing appears more in contradiction with the technology of behavior, as developed by Skinner*, denying the existence of the *Homo autonomous*, than the psychic phenomenology of Jung, or the mentalistic theory of the acquisition of *linguistic competence* of the adepts of Chomsky. Our function here is not to range ourselves in either line of thought, but only in stressing the existence of such a duality between concrete and abstract.

Note IV

After writing this chapter, we found in Fromm (*The Forgotten Language*) some concepts similar to those presented above. Freed of all interpretations by the followers of Freud and Jung, the language of the unconscious may have a deeper interpretation than to mean the satisfaction of repressed wishes or of forecasting what is going to happen to the dreaming individual in his immediate future.

Both interpretations are obviously biased and originate from aprioristical viewpoints of the founders of the psychoanalytical schools: Freud and Jung.

The language of the dream with their own codes and a different logic from that of the consciousness (that is that of the computer, as already mentioned), with their enigmas or puzzles, sometimes undecipherable by the dreamer himself, could constitute the primitive language of the living matter in its evolution, proposing problems that should be solved, as real puzzles, not very much different from those that the protoplasma must solve in the course of life to adapt itself to the new conditions of life in the surface of the planet.

Note V

We may mention also the "sonar" of the bats and dolphins, a system of emission and capture (reception) of ultra-sounds to orient themselves in the dark, to deviate from obstacles or to localize their preys (insects or fish), in an analogous way that the Englishmen used radar in the defense of London against the Nazi planes. Still more extraordinary is the adaptation of a moth receiving the ultra-sound emitted by a bat, through nervous receptors transmitting orders (electrical impulses) to its locomotor apparatus to deviate from the bat, by a 90° deflexion. *See* K.D. Roeder (1966) *Science* 154, 1515, and Idem (1965). *Scient. Amer.* 212, 74.

*B.F. Skinner (1972).

Note VI

The problem of pollution in the modern industrial world, has two aspects that are not always seen by those who endeavor to correct it indiscriminately. There is, in the first place, the "essential pollution" that is inherent to the industrial production proper, and the suppression of it, though making life more agreeable to the survivors, could provoke hunger and death of the thousands of individuals that depend upon it for their subsistence. It may be corrected, at long intervals of time, with the replanification of the industries, by recycling them and destruction of the polluting residues, or even by substitution of the raw materials by others that do not leave residues or that are easily degraded or eliminated with safety.

The correction of this type of pollution, will depend on the technological development or on severe impositions by the public powers. The second type of pollution is that which we could call "superflous," or even "*de luxe*," that consist of using (employing) difficult degradation products, as plastics, metal containers, gasoline of high explosive power (high index of octane) etc., to increase the attraction of consumers in a society as ours who prefer to receive the milk in plastic wrappers (bags), instead of the old glass bottle, or the beer in metallic containers, which give a peculiar click when opened on picnics, in public places and public gardens, leaving behind a metallic trail of difficult destruction; the excess of insecticides and fertilizers that aim to increase by no more than ten percent the capitalistic profits, and finally the immoderate use of tourist cars using gasoline of high pollutent power, to transport seventy to eighty Kg of individuals inside a carcass of two or three tons, sometimes at distances of two or three blocks in very heavy traffic, with hundreds of thousands of similar cars all contributing to increase the pollution of the streets in cities like Los Angeles, México City, São Paulo and so forth.

Some of these problems could be solved by returning to the useful glass containers returned after use; with a large increase of the cost of gasoline, as done in Brazil, where one liter of gasoline costs one dollar (65 cruzeiros). In such a way, only the very rich could pay for such an expensive means of transportation, but that would pay for the pollution that is produced with an over-tax that should be used to cover the increase of the collective transport, the price of the diesel, and the electricity.

Finally, to limit the use of fertilizers to levels that are sufficient to recover the fertility of the ground. Some of such problems only appear in the overpopulated centers, as New York, Los Angeles, São Paulo, but others, as the excess of insecticides and fertilizers, can affect the biological equilibrium in an irreparable way, on a universal scale. A simple presentation of the problem can be found in B. Commoner (1972), *The Closing Circle: Nature, Man and Technology*, Bantam Edit., N.Y.

Chapter IV

THE ORIGINS OF SCIENTIFIC EXPLANATION: THE GRAMMAR OF SCIENCE

IV

THE ORIGINS OF SCIENTIFIC EXPLANATION: THE GRAMMAR OF SCIENCE

The objects of science, as those of language, are things which exist in the Universe.* Such things can be classified in two large groups: concrete and abstract. Note that we are not subdiving things in abstract nouns and concrete nouns, but into *things* that are abstract, and *things* that are concrete. The verbs to *think, to suffer, to hate, to love,* but also particles such as *and, or, to, from,* the articles, definite *the,* indefinite *a*; or the pronouns: *I, you, he, them,* are abstract things, though they may serve as liaison or modulation of concrete ones. When I say "Peter *or* Paul" I am establishing a connection between proper names that indicate concrete things, through a particle with an abstract or mentalistic meaning, because the existence of Peter *or* Paul entails the idea that the one *or* the other are being excluded of an action that may be concrete or abstract: "Peter *or* Paul may be chosen;" "Peter *or* Paul will be eating the last portion of the pudding." Though everything else may be concrete in the last sentence, the alternative *or* refers to a mentalistic option therefore an abstract one. The same could be said of the indicative "*this* book" in which, while the book is concrete, the indicative *this* presents an optional attitude (mentalistic) that opposes this book to any other that may be at sight or reach of the locutor.

This opposition between concrete and abstract may be at variance with Russell's idea of different languages, a substantive against a predicative one, though in reality the great opposition between human languages is referred to the opposition between objective and subjective, concrete and abstract, irrespective of the fact that such an opposition refers to substantives (nouns), adjectives, verbs, adverbs, pronouns, conjunctions or prepositions.

This categorization involves the eternal opposition between external and internal, body and soul, matter and spirit, behaviourism and mentalism,

---

*Paraphrase of Wittgenstein's initial aphorism: "The world is the totality of all facts, and not of things." *Tractatus logico-philosophicus.*

empiricism and idealism, and all the multiple oppositions that were placed at the basis of philosophical systems, or that were legated through written words, from the time men could register their achievements or anxieties in flat pieces of wood, ceramics, papyrus, parchment or paper or in the form of works of art, that came to us in the caves of the paleolithic, as messages from prehistoric man.

Let us fix that point that seems fundamental in the study of the origins of language a few million years ago, at the same time as man acquired (developed) the capacity to manipulate tools and became able to establish a clear distinction between concrete things such as stones, bones, trees, animals and other men or women, with which he would collide in the outside world; but also describe concrete actions such as to beat, to eat, to drink, to run, to break, to shout, to speak, and adjectives such as colorful, blue, green, red, yellow, large, small, luminous, opaque, transparent and so-forth; in contrast to abstract nouns denoting abstract things, verbs that denote mental actions, adjectives that indicate purely mentalistic judgments, as formidable, magnificent, miserable, and all those already mentioned which can be derived of abstract nouns, as well as conjunctions and disjunctive particles, that in their essence denote mentalistic options about concrete and abstract *things* that are the objects of science and common language.

There is a point that we have to stress in relation to language, that will allow us to expand its analogy with science, namely that it is not enough to line up words (phonemes) with concrete or abstract meanings. Every student of first degrees knows or must know that all words in a phrase must obey certain grammatical rules, though the meaning (semantics) may remain indeterminate or arbitrary. The rules of syntactics or morphology of a language must be obeyed in order to generate *sense*.

We have seen that the latter (semantics) is more dependent upon the former (morphology and syntactics) than these upon the sense proper. The phrase can be of such an order that the sense will depend on the interpretation (judgment) of the hearer (as in poetry, for instance), though on the contrary a phrase completely destitute of grammar will be understood hardly at all or only exceptionally by the hearer as semantically acceptable.

Such a conclusion, rather simplistic or even unacceptable by many philosophers and linguists *a outrance,* is enough to allow us to extend the analogy with science. The truthfulness of a phrase or its acceptability rests upon the fact of everyday observation that even the illiterate or a minimally educated individual feels shocked, sketching a smile of superiority when he hears a phrase obviously wrong from the phonetic or syntactic point of view, though the somehow distorted meaning can still be understood or at least guessed by the hearer. In accordance with the ideas of Chomsky and his colleagues, one could say that the "deep structure" (semantics) of the phrase may preserve a certain character of invariance in relation to the

CHAPTER IV

superficial (phonetic and syntactic) structure, provided the alterations of the phrase (its grammaticality) are not of such a magnitude to render impossible the decodification by the hearer, i.e., the individual to which one speaks. The child who learned his mother tongue in surroundings of good cultural level will relish the mistakes committed by adults or foreigners who do not use with perfection the instrument that for him is already "*his mother tongue.*"

The situation is very different in the creative arts, in music, and possibly also in poetry, in which the artist feels totally noncommitted to rules and laws to be interpreted semantically by the customers of the art, that is, the appreciator of the work of art, the public in general, the one that receives the final product and must interpret it according to its mentalistic taste. In that case, the "meaning" does not depend so much of pre-established laws, at least as far as modern art is concerned.

Anything that the artist decides to be valid in his art can be interpreted as valid by the art consumer. We may say that modern art is a game without rules, in which anything can be worthwhile (or goes) if both the artist and his sustainer, (the art consumer) agree to complement each other, one to produce and the other to buy. On the contrary, language, in the sense of Wittgenstein,* is a game that obeys rules, without which language cannot be understood; in other words, "grammaticality" is a function that allows the hearer of any language to extract from any utterance of the speaker a meaning, an order, an information about concrete or abstract things in the universe.

Grammar, defined as a code or the rules of the game permitting to generate "an infinite or indefinite" number of expressions from a finite number of words (morphemes) or, from an even more restrict number of phonemes (three or four dozens in the known languages), it is, in the connotation of the transformationalistic grammarians, the instrument (tool) with which the users (speaker or hearer) become acquainted with its surroundings, using concrete and abstract names incorporated into an abstract or mentalistic not to say idealistic structure functioning as the most powerful instrument to be used by *Homo sapiens* to adapt its mental structure to "things" occurring in the universe. (*See* Notes I, II and III.)

In the case of science, it also deals with concrete and abstract *things,* as those that constitute the object of language. They can even be the same as in the case of psychology, and human sciences, but exceptionally the scientist would be interested to emit phrases that only concern concrete and abstract things of common language. But even in such exceptional cases, the "scientific" appears when the expression or phrase evolves from the common language to a more sophisticated one:

a. The *table* is made of wood.
b. The *table* is constituted of ligneous fibers.

---
*Wittgenstein, L., *Le cahier bleu et le cahier brun*, Gallimard, Paris, 1965.

c. The *table* is made up of organic matter derived from ligneous plants.
d. The *table* is formed of organic matter, containing mainly carbon, hydrogen, nitrogen and phosphorus;
e. The *table* is constituted of protons and electrons; and so forth.

The first phrase (a) contains only concrete things (table, wood, made), connected by abstract particles (is, the, of) that are part of the mentalistic common language that could be understood by the primitive man some few million years ago, when he could already use *his* articulate language to describe things of the universe in which he lived.

The second phrase (b), though still of only a descriptive character presupposes the knowledge of a wooden texture formed by fibers that can be evidenced by their plans of cleavage evidenced by the use of primitive implements (a stone knife) or his own nails, and therefore this phrase could still be part of the repertoire of the primitive man, with some adaptations to the language of his times.

The third phrase presupposes already a capacity of scientific classification of the materials existing in the universe, as inorganic (or mineral) and organic, as derived from living bodies, what became explicit in a much further stage of the development of the *Homo faber*. Under certain aspects, phrase (c) is much more colloquial than phrase (d), that presupposes a more profound knowledge of chemistry, that became completely explicit only at the end of eighteenth and beginning of nineteenth century, and even today would be an expression alien to the common man, needing a brief digression to make it clear about what is meant by organic matter, in the scientific nomenclature.

Finally, the last phrase (e) would be completely incomprehensible to man, even the scientist, at any time during the nineteenth century or beginning of the twentieth century, and only became semantically acceptable in the second decade of our century, but even today would be totally unintelligable to the common man without a college education.

But, let us analyze in a deeper way the last phrase (e), according to the rules of formation of scientific concepts. It would be very unusual for a scientist to be interested in the study of the intimate structure of his working desk (or table). For him, table in the expression given in (e), is only an example of another material object, such as a notebook, a textbook, a pencil, a piece of paper, and in his transformation grammar, any "concrete thing" may take the place of *a table* and, as a more "scientific" alternative, he would look in his lexicon for a word something like "matter," and would formulate the scientific expression:

f. Matter is constituted of atoms; or
g. Matter is formed of elementary particles (protons and electrons).

and so forth. And so would go on to translate into the "scientific" language, the initial, colloquial and grammatically correct: a) "The table is made of wood."

But in his eagerness to complete his hierarchic arboreous acheme (*phrase marker,* or *syntagmatic index*) the scientist would take each one of the components of his phrase:

h. The matter is constituted of elementary particles (protons, electrons, neutrons) and start his work of grammatical transformation to qualify the diverse components of the phrase rewriting: "the *proton* contains a positive charge, maintained in the nuclear structure associated to *neutrons* of equal weight but destitute of electric charge; an *electron* is a particle with a negative charge equal but of opposite signal as that of the *proton,* and animated by movement in a circular or elliptic orbit, to finally reach an expression with high scientific flavor, in:

i. The matter is constituted of protons with a positive charge (+), associated to a chargeless neutron (*o*), forming the nucleus, around which gravitate electrons with a negative charge (-) with weight (*w*) about 1.800 times smaller than that of the proton (*wwww*) or of the neutron (*wwww*): or else:

j. Matter is formed of nuclei (formed by protons and an equal number of neutrons) with a positive charge (+), around which gravitate electrons, with an equal negative charge (-), and weight (*w*) very much inferior to that of the proton (*wwww*) or the neutron (*wwww*), and so forth.

The generality of such an expression is of such an order that we can apply it not only to the objects used by man, as to all elements existing in the universe (with the exception of the hydrogen atom itself), which means that everything that has been found in the face of the earth, the planets, the sun and the stars are sufficiently covered by such a phrase. It is worth noting that inspite of the complexity of that phrase it is nothing more than the result of an analysis in scientific language of the elements contained in the Mendeleev Table.

We know that only hydrogen does not agree with such a description because in its natural form the nucleus does not contain a neutron, but only one proton, but there is always the possibility of considering its isotopic forms, deuterium with a proton and one neutron, and tritium with one proton and two neutrons. Everything else existing in the universe, from helium on, will answer to that "semantic formula," using only concrete things that could be objectively demonstrated by chemical or spectral analysis: protons, neutrons and electrons, each one of them answering the specifications or indications cf the lexical matrix in Table I.

With such concrete and abstract notions we can build up everything that exists in the universe, the hundred and four natural and artificial elements of the Mendeleev Table, and everything the components of which depend only of physical-chemical interactions as those existing in the living body, in the ionization of gases, and the cohesion of natural products in the face of the earth, and everything that concerns our everyday life. If the scientist

Table I

The lexicon to be used in such transformation

proton	charge (+)	weight (*wwww*)	spin (⇌)	magnetic momentum	(±)
neutron	charge (o)	weight (*wwww*)	spin (*o*)	magn. moment.	(o)
electron	charge (−)	weight (*w*)	spin (⇌)	magn. moment.	(±)
particle	charge (o,±)	weight (*w* to *wwww*)	spin (0, ⇌)	magn. moment.	(o,±)

would behave as the naturalists who founded natural sciences in previous centuries, collecting animals and plants and giving them scientific names, as Linnaeus with the binary nomenclature, Cuvier with comparative anatomy, Leuwenhoeck with protozoa, everything would be solved with the elucidation of the constitution of matter, with protons, neutrons and electrons, with a further introduction of other elementary particles, neutrinos, anti-protons, positive electrons, mesons, antimateria, etc, to continue the morphological description of the universe. All these semantic elements would enter in the matrix indicated in Table I and science would be no more than an immense morphological description of the three kingdoms of nature.

However, the transformational grammar of science continues its creative work, and already in the setting of such morphological elements in Mendeleev's Table, it was necessary to introduce abstract concepts that started to make their triumphal entrance with the quantum idea about emission and absorption of radiation by the blackbody, in the year 1901, by Max Planck* and the stationary orbits in Bohr's atomic model.

Let us make this point clearer. The electron is a concrete "thing" that can be "seen" by relatively simple means or apparatus as Geiger counters, for instance. At least it is more concrete than love, honor, dignity, and so forth, but its stationary orbit in the hydrogen atom is an abstract concept, as is abstract the concept of magnetic momentum, spin, field, etc. We know that they are abstract because their existence depends upon a mentalistic construct based in mathematical deductions.

To understand this viewpoint it is enough to consider how the concept of stationary orbit has been developed by Bohr, in 1913† in his famous atomic model based on the quantum concept of emission and absorption of radiation through entire *quanta* of energy. Previously, the atom of hydrogen was visualized as formed of a nucleus with a positive (+) charge,

---

*Planck, M., "*Uber das Gesetz der Energieverteilung im Normalspektrum*," Ann. Physik, 1901, 4, 553.

†Bohr, N., *On the constitution of atoms and molecules*, Phil. Mag., 1913, 26.1

surrounded by an elementary particle, the electron with a negative (−) charge, in the model described by Rutherford in 1911, the most popular one by that time.* The semantic problem (deep structure) was that of representing a charged (−) particle with a great angular velocity, behaving therefore as a periodic micro-current that would emit electromagnetic radiation (in the form of waves, the length of which would depend upon the dimensions of the orbit). Obviously, the emission of such an oscillating movement would forcibly lead to a rapid dissipation of the kinetic energy of the electron, and therefore describe a suicidal loop and fall down in the proton (+).

Anyone could say in simple colloquial language: "Under such conditions the hydrogen atom, and *a fortiori* any other element built up according to Rutherford's model, could not exist in any stable condition," and the emission of radiation would form a continuous spectrum in frank disagreement with the experiment that shows the hydrogen spectrum formed by discrete lines resulting from emission of radiant energy at delineated bands of frequency. The only way of stabilizing the hydrogen atom would be to assume (or *decree*) that while in orbit, the electron would be *forbidden* to emit radiation.

The next step taken by Bohr was that of transferring such a purely semantic, "mentalistic" concept of the stability of the "stationary orbit" of the electron into scientific language, using some transformational rules to *rewrite* the components of his initial phrase, "to stabilize the hydrogen atom, while in its orbit, the electron is forbidden to emit radiation." This transformation proposed by Bohr, used the concepts of Newtonian force and electrostatic attraction to define what could be the first stationary orbit of the electron in the hydrogen atom, simply assuming that the centrifugal force (mv^2/r) must equilibrate exactly the electrostatic attraction that, according to Coulomb's law would be proportional to the product of charges (e^2) to the inverse of the square of the distance (r^2). But marvel! That was exactly Rutherford's model that had been shown to be inviable in accordance with the laws of emission and absorption of radiation; therefore, it should be in the second point, namely in the suicidal continuous emission of radiation that would intervene the model proposed by Bohr. If the orbit *is* or *must be* stationary then the emission is impossible, and then such an emission can only occur when the electron *jumps* from one orbit to another of less energy. Inversely, if energy is absorbed, the *electron must jump* to a more external orbit, located further from the nucleus.

The great innovation of Bohr has been to label each orbit with a number (n) the square of which would be proportional to the dimensions of the stationary orbits departing from value $n = 1$, corresponding to the deepest

*Rutherford, E., 1909. *See also* cit., in A.B. and pg 75.

(ground state) orbit of the hydrogen atom. These numbers (principal quantum number) would have the values of the sequence of the natural digits 1, 2, 3, 4... and each time the electron jumps from one orbit *(i)* to another *(k)* it must emit or absorb an entire quantum of energy that would conform to Planck - Einstein law E = hv.* Some comfusion arises about the order (or senquence) of these digits. Since the potential of the electron in each orbit decreases as it approaches the nucleus, and is maximum when it is free from its bindings with the atom, it is more natural to consider the maximum as the one corresponding to the free electron, and the sequence given in the order 1, 2, 3,...∞ as the order of decreasing potential down to 0 when it is nearest to the nucleus.

To each stationary orbit corresponds a level of energy E_i (i = 1, 2, 3,...∞) inversely proportional to the square of n (n^2 = 1, 4, 9, 16,...∞). To each quantic jump corresponds the absorption or emission of a monochromatic radiation of frequency v_{ij} to be calculated by the general equation:

$$v_{ij} = \pm (E_i - E_j)/h$$

Theoretically, it is possible to use all combinations of digits (n = 1, 2, 3...) with values given to i, j in the above equation. For instance, in a quantic jump from orbit (n = 1) to orbit (n = 2) the above expression for the frequency $v_{21} = (E_2 - E_1)/h$, and so forth. The great innovation introduced by Bohr was that of establishing a one to one relation between the expression of emitted frequency and the quantic numbers n_1 and n_2 which characterizes each orbit, that of *depart* and that of *arrival,* or vice versa. The perfect reversibility of the equation:

$$v_{ij} = \pm (E_i - E_j)/h$$

may suggest that the phenomenon is not merely a physical one, but that some geometrical pattern may lie below the mechanism of emission and absorption of radiation by the electron when it jumps from orbit 2 to 1 or vice versa, and that some geometrical considerations might regulate such jumps. (*See* Note V.)

Continuing our brief story of Bohr's atom, there was a known fact that under normal conditions, the radiation of the H atom attains a maximum of frequency expressed by the famous Rydberg's constant (R = $3,290.10^{15}$ cycles per second), as indicated in Bohr's paper, referenced earlier. This value corresponded to the emission by an electron that changes its status from a free particle to an enslaved one in its ground position (deepest orbit in the H atom). Calculated in terms of energy it would also be equivalent to the so called "ionization potential" (= - 13,6 electron volts) of the hydrogen atom. All other frequencies emitted constituting the emission spectrum of

*Einstein, A., *Ann der Physik*, 1905, 17, 132.

the H atom could be calculated by the known formula proposed by Balmer (1885) and completed by Rydberg (*See* Karplus, M. and Porter, R.N. 1971):

$$v_{ij} = R \left(\frac{1}{n_i^2} - \frac{1}{n_j^2}\right)$$

in which R expresses Rydberg's constant in terms of frequency (cycles per second), and n_i and n_j are integers obeying the condition $n_j = m_i + 1$; $n_i + 2$; $n_i + 3$...etc. In accordance with Bohr's model, the constant in terms of frequency (cycles per second) would be calculated in terms of $e = 4,7.10^{-10}$ (charge of the electron); $e/m = 5.31.10^{-17}$ and $h = 6.5.10^{-27}$, (Planck's constant), by the formula

$$\frac{2\pi^2 me^4}{h^3} = 3,1.10^{15} \quad \text{cycles per second}$$

that corresponded to the usual value given in the literature as indicated above.

Nowadays, Rydberg's constant (R_H) is given in wave numbers $R_H = 109,677$ cm^{-1} that can be transformed in the above frequency by multiplying by the light velocity (c = 299,792.462 m/sec):

$$v_{max} = R_H c = 3,287.10^{15} \quad \text{cycles per second}$$

Here the great heuristic coincidence occurred that made the mentalistic representation of Bohr's stationary orbits acceptable by the most stubborn empiricists. Note that the agreement between Bohr's theoretical value and the accepted experimental one was not altogether perfect, but it was of such a magnitude to induce human brain to accept such a revolutionary concept as that of stationary orbit, not without some reluctancy. Such a concept was commented by Rutherford in a private letter (March 20, 1913) to Bohr: "*Your ideas as to the mode of origin of spectrum of hydrogen are very ingenious and seem to work out well; but the mixture of Planck's ideas with the old mechanics make it very difficult to form a physical idea of what is the basis of it. There appears to me one great difficulty in your hypothesis, which I have no doubt you fully realise, how does an electron decide what frequency it is going to vibrate at when it passes from one stationary state to the other? It seems to me that you would have to assume that the electron knows beforehand where it is going to stop.*"*

---

*N. Bohr., *Reminiscences of the founder of nuclear science and of some developments based on his work*, 1958. Lecture given at the Physical Society of London, Nov. 28, 1958. Published in *Rutherford at Manchester*. Edit. by J.B. Birks, W.A. Benjamin Inc., 1963, New York, pp. 114-167. In this same book the other cited papers by Bohr and Rutherford, are also found.

Other parameters of the Hydrogen atom could be found from pure dynamic concepts of the stationary orbit (such as the radius of the deepest orbit: $a_o = 0.5$ Å, or 5.10^{-9} cm that could be calculated by an equation that could be deduced entirely from a combination of classical mechanics and the old quantum theory:

$$r = \frac{h^2}{4\pi^2 me^2} \cdot n^2$$

where it is clear that the radius increases with the square of the quantic number (n).

The other elements of Mendeleev's Table could be built up using such a phoneme of the scientific language, by applying the needed transformations as the *Aufbau* principle, a sort of code of grammatical rules used to generate the elements existing in the universe.

Further transformations became possible by the introduction of much larger generalities, constituting the basis of the "new quantic mechanics." or "wave mechanics" following the work by De Broglie, Schrödinger, Heisenberg, Dirac, Pauli, and many others. This led to the idea of indetermination and the statistical foundations of prevision in modern physics, what we may call the "peak of irrationality" in the prevision of events in our physical or physical-chemical world. This phase was commented upon by Einstein who said "God would not play tricks with us," implying that a more rational or deterministic concept should be found to forecast events instead of merely playing dice in that kind of lottery, or hazardous gambling with nature (*See* Note V.)

But is that legitimate to consider all events in the universe, indeterminate or resulting of a hazardous gambling? It is clear that the velocity of light and the spectral lines of frequencies can be determined with such an accuracy unbelievable in not too remote times*. It appears obvious that indeterminacy belongs to space (in which Planck's constant $h/2\pi$ plays a predominant role, for instance in transforming "time mass" into "space mass"). But every phenomenon depending only on time (or frequency) can be measured within the utmost precision. (*See* Note V)

But, let us go back to that unique moment in the history of physics, which took its origin in the first year of this century with the quantic theory of emission of radiation by a blackbody, as formulated by Max Planck, in 1901. For a few years, the work of Planck introducing quantum theory remained relegated to a secondary position, as an empiric (*ad hoc*) solution for an isolated natural event, as that of emission of the blackbody until Einstein emitted his theory about the photoelectric phenomenon (Frank

---

*See A.L. Schawlow, "Laser spectroscopy of atoms and Molecules," *Science*, 1978, 202, 141-156.

and Hertz' effect), that gave civil rights to the light corpuscle (the photon), as a rational explanation to the strange law of quantic emission of the blackbody. And finally all that prepared the theoretical background to Bohr to emit his theory of the "stationary orbits," which by a pass of magic stabilized the hydrogen atom from the suicidal loop inevitable in Rutherford's model, at the same time giving a deep meaning (semantic) to Planck's theory. We may say that Planck stabilized the emitted radiation (avoiding the *Ultraviolet catastrophe*), that Einstein stabilized radiation with the idea of light corpuscles (photons), and that Bohr avoided the collapse of Rutherford's model with the concept of stationary orbits and quantic numbers. Finally stationary orbits were explained if not stabilized, by Schrödinger equation. However we may think of another way of stabilizing stationary orbits, as we are going to see later, by depositing the electrons in geodesic lines of a non-Euclidian space-time continuum. And since the metric of the space can be entirely defined by a bi-tensor meaning the total kinetic energy of the particle, changes in metric can only occur by emission or absorption of a quantum of energy (frequency × Planck's constant). (*See* M. Rocha e Silva. 1978. *S.S.T.* Note V.)

What appears of the utmost importance from the point of view of the grammar of science is the transformation of the electron and atom of hydrogen, introducing new specifications to amplify the meaning, i.e., the semantic of the expression: "The atom of hydrogen is formed by a proton surrounded by an electron." Though obeying still the equations of classical mechanics, important innovations were introduced in the definition of the hydrogen atom:

1. The electron occupies stationary orbits.

2. While in a stationary orbit, the electron does not emit or absorb radiations.

3. The stationary orbit is numbered by the series of integers departing from $n = 1$.

4. The transit from one to another of the numbered orbits, can only be done by emission or absorption of an entire quantum of radiation; the value of the energy ($E_{1,k}$) of the quantum is determined by the frequency ($v_{i,k}$) of the emitted radiation through the universal Planck's constant ($\hbar = h/2\pi = 6.5.10^{-27}$); according to Ritz' combination principle, the emitted or absorbed radiation has a frequency $v_{ij} = v_i - v_j$. If the electron jumps from an orbit i to k, it emits a monochromatic radiation, the frequency of which is $v_i - v_j$ represented by the symbol v_{ij}, and according to Ritz' principle all combinations of the digits i or j (i and j) give a possible frequency $v_{i,j}$, corresponding to the different levels of energy (E_1, E_2, E_3 ...) when multiplied by Planck's constant $\hbar = h/2\pi$.

5. The different values of energy (E_4, E_2, E_3...$E\infty$) of radiation will

determine the sequence of bands (lines) in the spectrum of hydrogen, according to the formulation of Rydberg, whose constant (R) is univocally related to other universal constants, the charge (e), the mass (m) of the electron, and Plank's constant (h);

6. According to the labels $n_1, n_2, n_3 \ldots$ of the orbits of departure and arrival, the series of Lyman ($n_i = 1$; $n_j = 2, 3, 4 \ldots$), of Balmer ($n_i = 2$, $n_j = 4, 5, 6 \ldots$), of Pashen ($n_i = 3$; $n_j = 4, 5, 6 \ldots$) and Brackett ($n_i = 4$; $n_j = 5, 6, 7 \ldots$) to be inserted in the formula of page 75 giving the values of emitted frequencies (ν_{ij}).

Therefore, in the transformational grammar of science, it was possible to expand the "deep meaning" of the expression, "the atom of hydrogen is formed by one proton and one electron," in a manner that one might describe if not complete, at least extensive, departing from a mentalistic explanation that might be called "artificial" or "stressed," by introduction of the concept of "stationary labelled orbits," based on a coincidence of theory and concrete phenomena observed in spectral analysis.

That such an explanation would require a more advanced (mentalistic) reformulation, or at least a more convincing one, was shown by the history of Physics in the subsequent years, when, in early twentieth century, De Broglie* with his "wave mechanics" and Schrödinger† with his famous equation, provided for the justification of the labelling quantum numbers proposed arbitrarily by Bohr, since 1913.

We have delineated above one of the methods of scientific discovery, leading to the expansion of the grammar of science, by combining the two entities in which the world is subdivided; namely, *the abstract and the concrete*. As human language constitutes a code to label things, using the abstract as a glue to establish deep relations (semantics) between things, we see that the hydrogen atom thought to be a mere indivisible corpuscle in the old conception of Democritus, a few hundred years B.C., expands itself in fields of force, centrifugal, electrostatic, gravitational, to become an association of wave and corpuscle, submitted to strange optical and dynamic influences (laws) hardly suspected by the most mentalistic minds of previous centuries.

We may distinguish in that story two distinct ways of arriving to the truth. First, we can see Bohr's way, in which the mentalistic image of the stationary orbits corresponding to discrete levels of energy ($E_1, E_2, E_3 \ldots$) suddenly becomes accepted by the most empirical minds, once a coincidence was proved between the theoretical value of Rydberg's constant and the experimental value known in the literature, since the great

---

*L. de Broglie, *Ondes et Mouvements*, Gauthier-Villars, Paris, 1926.

†E. Schrödinger, *Ann. der Physik*, 1926, 79, 361 and 489; 80, 489; 81, 109.

discovery of Balmer's simple formula to account for the lines of the visible spectrum. Once this coincidence was found, the whole theory fitted as a glove to the empirical, experimental data obtained by spectroscopy. This kind of discovery may be quoted as of the "hook-to-the-eye" type, as in the old-fashioned way, when tailors still used crochets as fasteners. The second way may be exemplified by the discovery of Schrodinger's equation, which followed immediately the discovery of De Broglie, of the wave-corpuscle; in that case, the adjustment was made in the continous zipper way, and in such a case, the discovery of the zipper was the consequence of the use of discrete crochets and possibly could not precede it, so the discovery of Schrodinger was the consequence of the slow transformation of abstract concepts starting with the old quantum theory (Planck, Einstein, Bohr) through the revolutionary idea of De Broglie of the wave-corpuscle leading to the *wave mechanics* or new quantum theory.

But let us go through this fascinating history that went through the years 1923-1930. Bohr's atomic model served as a mentalistic explanation until the appearance of a better explanation, in the twenties, with the introduction of wave mechanics. What was still unconvincing in Bohr's conception was the introduction of the quantic labels (digits, $n_1, n_2 \ldots$) in a somehow arbitrary manner, empirical or *ad hoc*. A scientific explanation needs something to satisfy human mentalistic appetite that tries to encircle or assimilate the universe. The introduction of a merely empirical parameter, though satisfactory to adjust the theory to experimental data of spectral analysis, was not mentalistically convincing.

Around 1923-1925, Louis de Broglie in his doctoral thesis, proposed a new image of the electron based obviously in the corpuscular nature of the photon, with a momentum (kinetic energy) and a monochromatic frequency (v), each other related through Planck's relation

$$E = hv$$

combined with Einstein's equation $E = mc^2$ would bring immediately to the magic relationship:

$$p = mc = hv/\lambda$$

establishing a relation between momentum ($p = mc$) and frequency

$$v = c/\lambda$$

Transfering such a fundamental relation to the electron in the hydrogen atom, Broglie created the wave or undulatory image that became the foundation of a new way to describe the elementary particle

$$n\lambda = hv/p = 2\pi r$$

that established in a simple way that in the stationary orbit the electron of

80 THE ORIGINS OF SCIENTIFIC EXPLANATION

radius (r) must fit into an entire number of wave lengths ($n\lambda$). This relation completed the mentalistic model of the electron as a particle of mass m, to which was associated a wave length λ. This notion extended to the proton, created the image for matter that was typical of light, the "wave corpuscle" of a frequency v, associated to a corpuscle of mass (m).

What was gained in satisfaction in the new explanation by Broglie was the rationalization of the quantum number that must now adjust exactly to the wave movement of the electron in the following scheme:

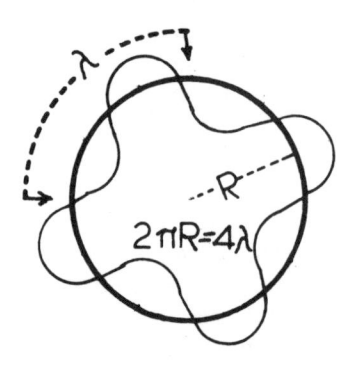

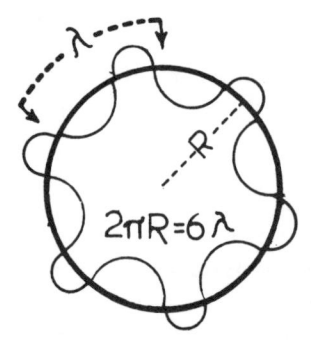

in which the wave movement of a wave length λ and frequency $v = c/\lambda$ must adjust itself to a circular orbit ($2\pi r$) by an entire number (n) of lengths (λ). Plank's constant (h) appears as a new constant of proportionality between mass (m) and frequency (v), in accordance to the fundamental equation

$$E = hv = mc^2$$

From this relation a series of semantic consequences permit to be calculated: frequency, in terms of mass or energy (electron-volts, joules, calories, etc.) as it may be found in more specialized books on the subject.*

The important aspect for the grammar of science is the way by which matter had been dematerialized in Einstein's equation transforming itself in vibration with a certain wave length (λ) and frequency (v), and its position *in space* determined by a probabilistic function (ϕ). One may argue that the gain in formalism was not very great because the same facts that were explained by Bohr's model could be explained perhaps in a more convincing way by the wave-corpuscle conception of Broglie. In reality, however, the gain has been immense, if we consider that less than two years later the existence of a wave associated with the electron could explain

---

*M.W. Hanna, *Quantum Mechanics in Chemistry*, 2nd edit. W.A. Benjamin Inc., 1964, New York. Also, J. Barrett, *Introduction to Atomic and Molecular Structure*, John Wiley & Sons, 1970, New York.

another phenomenon that would remain otherwise unaccounted for, the so-called bands of diffraction of the electron, in the famous experiments by Davisson and Germer (1927)* and Thomson (1927)†. It became then possible to measure the wave length ($\lambda = 10^{-8}$cm) and frequency ($\nu = c/\lambda = 3.10^{18}$) in good agreement with the previsions by De Broglie (1924). From that, to the invention of the electronic microscope was a short step, and then to the discovery of the protonic microscope. But more important than that, it introduced the idea that to any material particle of mass m is associated a wave of frequency (ν) in accordance with Broglie's fundamental equation and Planck-Einstein's relation

$$mc^2 = h\nu$$

If we had to select a musical about this heroic phase of physics, in the period 1913-1930, we might choose Scherazade of Rimski-Korsakov, in which Sindbad's boat is sailing in alternatively stormy or quiet seas, traveling along on such ethereal medium as the musical phrase. By acceptance of Broglie's wave theory the "hook-and-eye" theory (Bohr's model) became insufficient. We needed something more poetical or musical (a "zipper theory") to describe the electronic orbit, something like a violin chord, to support an entire number of semiwave lengths, to be adjusted to the whole circumference ($2\pi r$) of the orbit.

The problem to be solved in continuation was that of understanding how a corpuscle of mass (m) can exist inside of a train of waves of frequency (ν). But, more than that, it was imperative to know the nature of such undulatory movement, as postulated by Broglie, enslaved in a particle of any matter of mass (m). And here we must meet the mentalistic quality of the scientific theory. Those waves *were not* of an electromagnetic nature, but maintain with it the relation that all waves maintain with each other; i.e., obeying a mathematical equation establishing a relationship between the maximum value of the ordinate (amplitude), the frequency, and the period of vibration. Sound waves, light waves, waves in the surface of a lake or in a stormy or quiet seas where Sindbad's boat navigates, all of them obey a differential equation, the "wave equation" with the cabalistic form:

$$\frac{d^2}{dx^2} \phi = \frac{1}{v^2} \cdot \frac{d^2}{dt^2} \phi$$

where v is the velocity of propagation of the undulatory movement along the spatial dimension (x), in function of time (t).

Though formulated more than one hundred years ago from the undulatory movement of a vibrating string, this equation assumes a

*C. Davisson and L.H. Germer, *Nature*, 1927, 119, 558.

†G.P. Thomson, *Nature*, 1927, 120, 802.

semantic value in the grammar of science, and we could say that it is the algebraic translation of the verb *to vibrate, to wave, to undulate,* provided the vibration or the waving obey some other conditions that make them regular (sinusoidal). That means that the function ϕ to which the double derivation is applied can be expressed as a function of sine or cosine. Any function going up and down can be transformed in a sine or cosine function or superposition of such functions by applying the development rules of a Fourier series.*

The value of function ϕ, in any point of space or at any time of propagation, indicates the *amplitude*; in analogy with other undulatory movements, it indicates, in the case of the electron, a parameter of intensity, and its square ϕ^2, the probability of finding the electron in any point of space.

If the solution was that simple, any student would be able to solve the riddle of the quanta, as applied to Bohr's atom. The solution given by Schrödinger, in 1926 was a little more sophisticated, because the electron is not a simple undulatory movement, but instead, "a especial case" of an undulatory movement, with a mass (m), a charge (e), oscillating in a "discontinuous universe" regulated by Planck's constant (h), with a kinetic energy (U) and potential energy (V). In short, what had to spring from Schrödinger's brain was a new phoneme or syntactic rule in the grammar of science.

All things considered, Schrödinger (1926) proposed two kinds of mentalistic solutions, one "independent of time" (I) to account for the different levels (E_n) of energy of the "stationary orbits" or states:

$$\frac{h}{2\pi m} \left(\frac{d^2}{dx^2} + \ldots \right) \phi + (U - V) \phi = 0 \qquad (I)$$

And another (II) "time dependent:"

$$\frac{h}{2\pi m} \left(\frac{d^2}{dx^2} + \ldots \right) \phi + (U - V) \phi = i\hbar \frac{d}{dt} \phi \qquad (II)$$

Equation I can be put in a matricial form to make explicit the calculation of the different levels of energy (E_n):

$$[H] \phi = E_n \phi$$

where the matrix $[H]$ represents the total energy of the system ($T + V$), the so-called Hamiltonian H, operating upon the wave function (ϕ) (= $A \cdot \sin\alpha\, x + B \cdot \cos\alpha\, x$) reproducing the function (ϕ), multiplied by E_n, the different or possible levels (n) of the energy of the system.

*J.P.G. Richards and R.P. Williams, *Waves*, Penguin Books, London, 1972.

In a more technical language, if [H] is a matricial operator operating upon an *eigenfunction* (proper function)ϕ, it generates the *eigenvalues* (proper values) E_n (= $E_1, E_2,\ldots E_n$), integral multiples of a basic value E_o, and since these values are proportional to the squares of n, we will have $E_n = n^2 \cdot E_o$ this giving by a mentalistic process the justification for the quantic number (n) of Bohr's atom.

But now, the process of discovery was complete or almost complete. The quantic number is no longer an empirical (enforced), *ad hoc* one, but a necessity of the new model suggested for the electron as a "wave-corpuscle" entity. With the new concepts of wave mechanics, the principle of uncertainty formulated by Heisenberg in 1926-1927 according to which it is impossible to determine with equal accuracy (precision), the moment (velocity) and position of the electron in its intra-atomic or "secular existence," and the concept of *orbit* must be modified to that of an "orbital," i.e., an *electronic cloud* around the nucleus and its density in each point or circumference of radius R, can be measured by the square of the wave function (ϕ^2). In other words, in its precarious existence, the electron may be found in any region around the nucleus, though the maximum probability to find it is that corresponding to the old quantum theory, namely at a distance of the nucleus, corresponding to Bohr's atom, with a $R_o = 0.5$ Å or 5.10^{-9} cm!

The usual way of presenting the principles of the new quantum mechanics is in the form of postulates deriving immediately from the theory. However, such principles presented as postulates are hardly understandable intuitively and the best is to refer here to their formulation in the book by Hanna (1964) cited earlier.

We reproduce here the first postulate giving us the definition to the wave function:

Postulate I
(a) *Any state of a dynamic system of N particles is described as fully as possible by a function ϕ ($q_1, q_2,\ldots q_{3N}, t$) such that:*
(b) *The quantity $\phi * \phi \, dt$ is proportional to the probability of finding q_1 between q_1 and $q_1 + dq_1$, q_2 between q_2 and $q_2 + dq_2 \ldots q_{3N}$ between q_{3N} and $q_{3N} + dq_{3N}$ at a specific time t.*

What this postulate says (in less succinct form) is that all information about the properties of a system are contained in a function (usually called a wave function) which is a function only of the coordinates of the N particles (with 3 degrees of freedom each, therefore with a total $3N$ d.f.) and the time t. If the wave function includes the time explicitly, it is a time-dependent wave function. If the observable properties of a system do not change with time, the system is said to be in a *stationary state*. A ϕ function describing such a state is called *a stationary state wave function*...The second part of the postulate gives a physical interpretation of the ϕ

function. This interpretation is easiest to visualize for a system containing a single particle constrained to move in one dimension. The quantity $\phi * \phi\, dx$ is then just the probability of finding the particle between x and $x + dx$ at a given time t. A ϕ function can be complex; hence the "probability density" $\phi * \phi$ is a product of ϕ with its complex conjugate $\phi*$. (Hanna, 1964, Loc. cit.)

Now, we are sailing again in quiet waters in Sindbad's ship. In our transformational grammar of science, the electron is now represented by a wave function and its properties are deduced from the postulates of wave mechanics. Its *orbital* (no longer orbits) around the proton, in the hydrogen atom, are mere solutions of Schrödinger's equation. Strange pictures arose for the orbitals $s, p, d\ldots$ as electronic clouds of increasing complexities departing from the ground state:

$1s, 2s, 2p, 3s, 3p, 3d\ldots$

arising as mentalistic enforcements of the postulates of the so-called *Aufbau* principle, generating a permanent monument to the genius of Mendeleev, the one who has seen or forecast everything in the last quarter of the nineteenth century. Everything, or almost everything.

We may remain on the solid ground of the proton, neutron and electron, which are enough to explain everything that happens in everyday life, in the living world, in organic and inorganic chemistry. The electron in its semi-abstract existence, possess charge, mass, spin, frequency, angular momentum, kinetic and potential energy. In its precarious citizenship, in a Universe in which it is king without a kingdom, must obey the drastic stipulations of Schrödinger equation, and in its secular behaviour must obey Heisenberg's uncertainty principle.* Hardly any one could imagine a more powerful monarch tied up to exigencies of laws that permit the permanence of the Universe. But can we be sure that it is him, the elementary particle of the universe, indivisible corresponding to the real concept of Democritus atom? But such a powerful monarch, could at last proclaim his independence from so many abstruse properties. One day he might come with a proclamation to "whom it may concern:" "I, the almighty sovereign of this realm of elementary particles, I am nothing but a bundle of frequency. What you see of me in the field of your microscopes or imagination is the sparkling view of my temporal being in a tridimensional spacial screen. I play my deterministic role by measuring time, with a precision to the 100.000.000th fraction of a second, and along that line I am really the most powerful monarch of the *Realm of Time*."

That is the new development in the grammar of science, to be told in Note V.

---

*W. Heisenberg, *Zeitschr. f. Physik*, 1927, 43, 172.

CHAPTER IV

Note I

A fully adequate grammar must assign to an infinite range of sentences a structural description indicating how this sentence is understood by the ideal speaker–hearer. This is the traditional problem of descriptive linguistics, and traditional grammars give a wealth of information concerning structural descriptions of sentences. However, valuable as they obviously are traditional grammars are deficient in that they leave unexpressed many of the basic regularities of the language with which they are concerned. This fact is particularly clear on the level of syntax, where no traditional or structuralist grammar goes beyond classification of particular examples to the stage of formation of generative rules on any significant scale. An analysis of the best existing grammars will quickly reveal that this is a defect of principle, not just a matter of empirical detail or logical preciseness. [N. Chomsky, *Aspects of the theory of syntax,* MIT Press, Cambridge, Mass., 1965—impression (1969).]

Note II

The creative aspects of human language have been always recognized and are part of the general principles that characterize a possible universal grammar (U.G.). However, as stressed by Chomsky (1965, loc. cit.) there has been always the tendency of linguists (and unfortunately of teachers of language) to separate the descriptive study of language from its creative aspects belonging to a distinctive science that is called psycholinguistics or pyschology in general. To many, there would be languages that are more appropriate for certain forms of thinking, and the most elucidating or schematic case is that expressed by Diderot (1751) when he stressed the qualities of French to formulate scientific thinking; i.e. useful and concrete thinking, in contrast with other old and modern languages that would be more appropriate for the literature and poetry: "*Le francais est fait pour instruire, éclairer et convaincre; le grec, le latin pour persuader, émouvoir et tromper; parlez le grec, le latin, l'italien au peuple, mais parlez le francais au sage*" (Diderot, 1751; cit. in Chomsky, 1965). A very simple argument could be opposed to such an assertion of Diderot: for many centuries and even today, Latin has been the language of science, and nowadays English constitutes the most common language of international scientific congresses. But, let us see how Chomsky (1965) define a generative (and transformational) grammar: "A system of rules that in some explicit and well defined way assigns structural descriptions to sentences. Obviously, every speaker of a language has mastered and internalized a gerative grammar that expresses his knowledge of his language. That is not to say that he is aware of the rules of the grammar or even that he can become aware of them, or that this statements about his intuitive knowledge of the language are necessarily accurate. Any interesting generative grammar will be dealing, for the most part, with mental processes that are far beyond the level of actual or even potential consciousness; furthermore, it is quite apparent that a speaker's reports and viewpoints about his behaviour and his competence may be in error...Similarly, a theory of visual perception would attempt to account for what a person actually sees and the mechanism that determines this, rather than his statements about what he sees, and why, though these statements may provide useful, in fact, compelling evidence for such a theory." (Chomsky, 1965, loc. cit.).

Note III

It belongs to the essence of generative or transformational grammar to analyze linguistic events upon a mentalistic background, which in accordance with Chomsky and his disciples and apostles (Katz and Postal, 1964) guarantees creativity in the use of natural language. The revolution introduced by Chomsky in the 1950s and 1960s aimed to demonstrate the insufficiency of structural analysis to characterize the grammar of any language, in a number necessarily finite and therefore able to be formulated, to generate an infinite number of phrases and articulations. For that, any phrase, however simple it is, must be analysed through a reformulation or a rewriting, that constitutes the basis of the transformational grammar. The whole scheme of the rules of rewriting can be represented by a tree diagram:

Phrase ⟶ Noun + verb; Noun ⟶ (Art.) Noun;
Verb ⟶ Aux. Modal ⟶ etc., etc..

giving way to what Chomsky called *deep structure,* containing the meaning (semantics) of the phrase. The latter is represented by a code, with all specifications needed to generate an infinity of *superficial structures* (phonetic and syntatic). Symbolically, a phrase Σ of generative grammar can be expressed in the form of a function

$$\Sigma = f(S.P)$$

in which S represents the *deep structure* (semantic) and P (Phonetic) corresponds to the *superficial structure*. The function of a grammar, therefore, would be to make it possible a derivation between meaning (semantics) and the code, up to a certain point, through syntatic constructs. In that sense, syntactic and phonetics constitute the *surface structure* of the phrase (Σ). However, Chomsky's theory is more ambitious and the main objective of the generative grammar, already expressed in his book on *Cartesian Grammar* (Chomsky, 1969), and developed in his more technical books on *Syntactic Structures* (1957) and *Aspects of Syntactic Structures* (1965), and more recently in *Questions of Semantics* (1972), and still more recently in *Reflexions on Language* (1975, 1977), is to find out the laws or universal rules (possible common to all languages, and especially to Indo-European ones) that might account to the infinite variety of sentences to be formulated by a native which handles with competence his maternal tongue.

If we consider human language the best instrument to describe the universe, the field of application of the grammar of any language coincides with the Universe itself, and it is in such sense that the sub-languages of science, in mathematics, physics, chemistry, biology, sociology and so forth, all that is part of the field of natural sciences, constitute metalanguages created by human invention to complete the common natural language that the individual learns in its earliest years (up to the fifth or sixth years of age) in life. In the same way that the natural common language has been studied in its morphological (phonemes and morphemes) and structural (syntax and syntagmatic grammar) and semantic aspects, and only much later, it was thought to create an explanatory (generative and transformational) one. Also science evolved from a mere descriptive or morphological classification of the living bodies (in classes, genders and species), of the elements (tables of the classification of elements that preceded that of Mendeleev), of the physical phenomena, with their static and dynamic aspects, in classical mechanics, in geometric optics, acoustics, and electricity and magnetism. Only in more recent times, did science evolve into a functional (deep) explanatory structure using human creativity based in abstract concepts permitting the genesis of a number practically indefinite of new formulations: the genetic code, the chemical mediation of living functions, embryology, quantum, wave and undulatory mechanics, relativity theory, *Aufbau* theory of the elements departing from simpler configurations of the hydrogen atom and elementary particles, of subparticles which are being described in increasing numbers, and so forth. We may say that analogous to what is being developed in the study of language, science possess also its transformational (and generative) grammar. Such is the aspect that might give a new direction to the philosophy of science, instead of the way that follows philosophy of science as practiced by professors of philosophy, disconnected from the essence of the scientific method.

See also N. Chomsky, *Language and Mind,* Hartcourt Brace, New York, 1968.

Note IV

An example of a linguistic treatment, and, we could add, of the transformational grammar of science, can be found in the article by Weisskopf (1975): "*Atoms, Mountains and Stars: A Study of Qualitative Physics.*"

"Almost all of the material phenomena which occur under terrestrial conditions are recognized as quantum mechanical consequences of the electric attraction between electrons and nuclei and of the gravitational attraction between massive objects. We should be able, therefore, to express all the relevant magnitudes which characterize the properties of matter in terms of the following six magnitudes: M, the mass of the proton, m and e are th mass and electrical charge of the electron, c, is the light velocity, G, is Newton's gravitational constant, and—most important—h

is the quantum of action. In addition we will use the number Z and the atomic weight A of the elements whose properties we study, since Ze determines the charge and AM the mass of the nucleus."

With such phonemes of the language of science, on the basis of fundamental principles, such as wave-particle of De Broglie, Schrodinger equation, Pauli's exclusion principle, and derived constants, as ionization energy (I), Rydberg's constant (Ry) and R the radius of the hydrogen atom, D, the energy of dissociation of the molecule and B, the binding energy of atoms and molecules in the solid and liquid states, we arrive at the following qualitative rules of *rewriting*.

$$I = \alpha\ Ry\ (1 > \alpha > 1/4)$$
$$R = f\ a_o\ (6 > f > 1)$$
$$D = \beta\ Ry\ (0.5 > \beta > 0.2)$$
$$B = \gamma\ Ry\ (0.3 > \gamma > 0.1)\ \text{solids}$$
$$(0.1 > \gamma > 0.05)\ \text{liquids}$$

It was possible to Weisskopf, in a most ingenious way, though only in a semi-quantitative (or qualitative) way to describe the relations betwen atomic nuclei, density and hardness of matter, the height of mountains, the velocity and wave length of the surface of a lake, the size and number of protons of a star. This article by V.P. Weisskopf in *Science,* 1975, 187, 605, is an excellent example of transformational or gerative *grammar of science* obeying the slogan: "*Science uses a finite number of rules (of principles, or laws) to describe an infinite number of events in Nature.*"

Note V

Attempts to recover rationality. One of the last phrases of Albert Einstein was a prophetic condemnation of the excess of irrationality of the physicists who were developing the new quantum theory, under the name of wave mechanics, obeying the postulates of De Broglie, Heisenberg, Schrodinger, and all the great physicists mentioned in the foregoing chapter: "*If the statistical quantum theory does not pretend to describe the individual system (and its development in time) completely, it appears unavoidable to look elsewhere for a complete description of the individual system; in doing so it would be clear from the very beginning that the elements of such description are not contained with the conceptual scheme of the statistical quantum theory. With this, one would admit that, in principle, this scheme could not serve as the basis of theoretical physics. Assuming the success of efforts to accomplish a complete description, the statistical quantum theory would, within the framework of future physics, take an approximately analogous position to the statistical mechanics within the framework of classical mechanics. I am*

CHAPTER IV

firmly convinced that the development of theoretical physics will be of this type; but the path will be lengthy and difficult." Einstein, Febuary 1, 1949 (In P.A. Schilpp, 1951).

This attitude of Einstein, was repeatedly remembered when his centerary of birth was commemorated in March 1979. The solution of such dilemma would depend upon finding a rational bridge between relativity theory and old or new quantum mecanics. In a series of papers published in a new and challenging periodical *"Speculations in Science and Technology"*(SST) we have developed a theory (Rocha e Silva, 1978-1979) on the emission or absorption of radiation by the hydrogen atom, by assuming that the electron is fixed in its stationary orbit, because it is bound to a geodesic in a Weyl's space-time continuum, and that emission or absorption of energy (E_n) involves a change of "gauge," along the relativistic time ($T_i^{-1} = v_i$) parameter. A few definitions were introduced to explain such adaptation of gauge when the electron jumps from orbit (i) to (k) or viceversa from (k) to (i), as a parallel displacement along the time coordinate ($T^{-1} = v$). This postulate is the simple formulation of the experimental fact (Ritz' combination principle) according to which the only parameter to identify the orbit is the frequency (v_i) of the particle. According to relativity postulates, the dynamic parameter g_{ii} was split in four components of mass: "spatial mass" (m_{11}, m_{22}, m_{33}) and "time mass" ($m_{44} = g_{44}$). The three components of spatial mass correspond to the values $m_{ii} = g_{ii}$ in Einstein relation $E_{ii} = m_{ii}c^2$ where

$$i = 1,2,3 \qquad (1)$$

while the time mass is given by the symmetrical relation

$$m_{44} = f_{ik}/c^2 \qquad (2)$$

where $f_{ik} = v_{ik}$ is the frequency emitted when the electron jumps from orbit k to i, according to Ritz' combination principle:

$$f_{ik} = \pm (f_i - f_k) \qquad (3)$$

If both sides of this equation are multiplied by $h/2\pi$, we have the expression in terms of energy:

$$E_{ik} = \pm (E_k - E_i) \qquad (4)$$

representing the transformation which introduced the Quantum indetermination, explicit in Heisenberg principle of uncertainty. It is easy to show that such a change of parameters from frequency (f_{ik}) into energy (E_{ik}) = $h/2\pi$. f_{ik} that was the revolution proposed by Bohr (1913) transforming a deterministic law (of frequencies) into the old quantum theory obeying the indetermination expressed by Heisenberg's uncertain principle. The suggestion contained in our papers (Rocha e Silva, 1978-1979) was to reverse the way followed by Planck, and consider the relativistic time parameter ($T_i^{-1} = f_i$) as the parameter holding the electron in its orbit (i). Such a transformation, permitted to calculate the main parameter of the orbit, using equation (3), with the new law of emission or absorption of radiation

$$\Delta m_{44} = \Delta f/c^2 \qquad (5)$$

reducing the process of emission or absorption of ratiation to a logarithmic function, according to the postulates of Weyl's geometry, as shown in Rocha e Silva (1978):

$$d(\log f) = \frac{\Delta f}{f} = \text{const.} \tag{6}$$

The consequencs of this law of radiation were presented in recent papers:
a) To reduce the *line spectrum of the hydrogen* to a phenomenon of relativistic "red-shift," with a coefficient of red-shift of the order of $\frac{\Delta f^n}{f} = 0.75^n$ where n is the principal quantum number; this enormous value for the red-shift coefficient produced by the electrostatic field of the hydrogen atom, is to be compared with coefficients of red-shift of the order of 10^{-28} of a gravitational field. Such an enormous value of the coefficient of the red-shift of the electrostatic field of the ionized hydrogen atom, induced one to postulate:
b) A generalization of the *relativistic equivalence principle* to apply to the electrostatic field of the ionized atom of the hydrogen, or of any ionized field produced by high temperatures, such that of the helium (fully ionized at temperatures of the order of 0.3 million degree °K centigrades.
c) Such a generalization of the relativistic equivalence principle to the electrostatic field, about 10^{48} times more potent than the gravitational field on earth, might account for the enormous red-shift at the surface of Quasars (equivalent to that of ionized hydrogen), to explain black-holes (as ionized helium masses), the proposed (or alleged) expansion of the Universe, big-bang, and so forth. This will be the object of separate papers (Rocha e Silva, 1979, 80.)

Chapter V

THE LANGUAGE OF LIVING MATTER

V
THE LANGUAGE OF LIVING MATTER

The biologist has no great difficulty in understanding that there is a language whose symbols and signs are printed in living matter just as the 10 commandments were printed in Moses' Tables by the natural forces of lightning, atop Mt. Sinai.

Living matter is sculptured by symbols represented by chemical groupings arranged in a linear, surface, or spatial order, that must be "deciphered" (read) by other structures, also belonging to the living matter. Often such a reading or decodification determines the execution of some important functions of the organism. When a certain enzyme attacks its substrate, releasing energy or originating metabolites of vital importance, it must initially "recognize" some chemical groupings existing in the substratum. On the other hand, the substrate in the innermost structures of cells and tissues, or when administered from the outside, must "recognize" the structure of the so-called "active center" of the enzyme, in order to suffer the "desired" transformation. We are using this teleological language purposefully because it is convenient, though it may be shocking to most radical empiricists.

In reality, the substrate must search for "its enzyme" just as the young fellow looks for his sweetheart; or the father for his son in an enraged multitude. But, a most suggestive comparison of the way the substrate looks for its enzyme, is that of the captain searching for a place to dock his ship by reading codes and signals coming from shore. The adjustment is then done between the chemical groupings of the "active center" of the enzyme and those that are peculiar to the molecule of the substrate.*

One deals here with a "reading," in the exact meaning of the word, of a sequence of symbols, the "phonetic" of which is represented by forms of

---

*For an extensive bibliography see: M. Rocha e Silva, 1973-1979, *Fundamentos de Farmacologia*, Edart, Livr. Edit, São Paulo. Translated into Italian and Spanish (1978). See also list of *Additional Bibliography (AB)*

attraction or repulsion between chemical groupings; and those are the forces that orient the molecules of the substrate to adapt themselves to a determined space limited by the units (chemical groups) existing in the macromolecule of the enzyme.

As mentioned above, the situation is not very much different from that encountered by the captain of a ship under a clear sky or in a dense fog attempting to orient his ship to board along the quay, by interpreting lights, sounds or telegraphic messages coming from shore, to find the place alloted to him.

One of the greatest problems of modern biochemistry is that of clarifying the nature of the forces that orient substrates to find the "active centers" destined to them in the enzymatic macromolecules. In congresses of biochemistry, or of biophysics or of pharmacology, one of the major attractions used to be the show offered by gigantic multicolor models of enzyme molecules, *digesting* or simply fixing up in a certain area (sometimes less than one percent of the total) of the macromolecule, its substrate, that is, the material over which it must exercise its metabolizing activity. Stereoscopic pictures to be observed with proper glasses, are already a matter of routine in modern textbooks of enzymology.

Such notions are new, very new, and represent the effort of thousands of biochemists working all over the world, trying to decipher such an enzymatic code. It is really fascinating that such a great number of human brains, each one with its 100 billions of neurons, are needed to realize in a few decades what the substrates in the intimacy of the cell do perform in some tenths or hundredths of a second. To the benefit of the former (the brains, each one with its 100 billions of neurons, are needed to realize in a few decades what the substrates in the intimacy of the cell do perform in such phonemes of the language of the living matter. Such is the "linguistic meaning" of the evolution of life on the earth's surface. (*See* Note I.)

Another example is the recognition of the molecule of an active agent (agonist) by the so-called "pharmacological receptors." Just as biochemistry tries to identify the "active center" of enzymes, modern pharmacology is based on the idea that the functions of the organism (synaptic transmission in the central, peripheral and autonomous nervous systems) are dependent upon the chemical stimuli of such relatively simple substances as acetylcholine; catecholamines (adrenaline and nor-adrenaline); polypeptides as bradykinin, angiotensin and substance-P and, more recently, prostaglandins, that are released or formed at strategic points of the organism, producing immediate or delayed hormonal actions. The normal functioning of the organism depends upon the interaction of "agonists" with their "specific receptors." About all products used in the medical practice, in a certain way, act through such receptors, inhibiting or potentializing the action of agonists or "chemical mediators" of physiological functions. Therefore, not only the chemical mediators (agonists) but also the substances that are used as medicaments must, to start with, recognize the sequences of chemical groupings (or radicals) existing in such receptors. And also here, teleologically, we must understand that the simple molecule of the agonist must search in the multitude of macromolecules of proteins, just that sequence that has been allotted to it in the effector organ.

A substance (drug) that *must* block the effect of acetylcholine in the parasympathetic nervous system or in the ganglionic synapses of the autonomous nervous system, or also in the synapses of the central and peripheral nervous system, must know how to read the linguistic code printed in the so-called post-synaptic membrane, in which the organism has labelled the sequence of atoms and chemical groupings existing in the molecule of acetylcholine, for instance.

This code or "morpheme" must be distinguishable from that destined to be recognized by nor-adrenaline (nor-epinephrine), histamine, serotonin or of polypeptides (angiotensin, bradykinin, substance P, endorphins, morphine analogues, etc.), prostaglandins, and much larger molecules, studied in endocrinology.

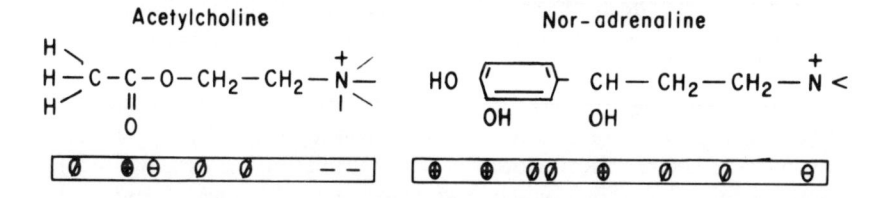

Ø, hydrophobic;   ⊕ ⊖, dipole;   —, ionic (weak);   — —, ionic (strong)

It is obvious that the importance of such recognition of symbols, used by the living substance to distinguish its chemical offsprings to avoid a certain function that depends upon the interaction of acetylcholine with the post-synaptic membrane, could be extended to other mediators such as histamine, nor-adrenaline, bradykinin, and so forth.

Still in this language of the receptors, we may admit varieties of symbols to recognize each one of the different functions exerted by the same mediator. So, acetylcholine acts as mediator of functions that may appear antagonist. At the same time that it is a mediator in the terminals of the parasympathetic nervous system (reduction of cardiac rhythm), stoppage of the heart, stimulation of the gastro-intestinal musculature, it may act also as mediator in the ganglia of the sympathetic nervous system, for instance, and in such a function its effects (acceleration of the heart, increase of arterial blood pressure, reduction of the movements of the stomach and intestine) are of course antagonical to the former. We face here a situation well-known to students of pharmacology, of how the organism is able to recognize in different manners the same mediator, what is made intuitive if one accepts the idea that the same mediator (ACH) has at its disposal distinct receptors, such as *muscarinic* and *nicotinic* receptors. (See Note III, page 107)

In the current human language it would be the case of synonym or homophone, in which the interlocutor must recognize the "meaning" by the context of the phrase or sentence. So a hearer of his mother language can easily distinguish the double meaning of "rico" (precious or wealthy) by its position in the phrase "o meu rico Português" (my precious Portuguese) from "meu Português rico" (my rich Portuguese), which is not very different from what the organism does to indicate to acetylcholine its function as a *muscarinic agent* or a *nicotinic one*, simply by orienting the sequences of the chemical groups in its molecule, in one direction or the opposite one. Therefore, when acetylcholine (ACH) must produce its muscarinic effect, it must recognize (or *read*) the sequence printed in the *muscarinic receptors*, which command the reduction of the cardiac rhythm and fall of arterial blood pressure, for other instances, the "intention" of th: organism may be to produce nicotinic effects (stimulation of ganglia or the transmission in the voluntary muscle end plates). In both cases, the messenger (mediator) is the same, namely molecular acetylcholine, sent to the right place at the right moment. The way this operation is produced is a mere question of dose and of the locale of production of the mediators. Small doses of acetylcholine (ACH) go directly to the muscarinic receptors, without affecting transmission in the ganglia or the transmission in the motor end plate. If larger doses are given, the effect upon the ganglia is produced but is masked by the muscarinic effects, which are stronger and antagonistic. The students of pharmacology know that to put into evidence the nicotinic effects (rise in blood pressure, acceleration of the heart) it is necessary to block the muscarinic effects with a previous injection of atropine and raise the given dose of acetylcholine. To show the effects on the motor end plate (striated muscle contraction) it is necessary to inject massive doses of the mediator (ACH) directly in the artery serving the muscle.*

The given names of *muscarinic* and *nicotinic* suggest that other substances (muscarine, extracted from the fungus *Amanita muscaria* and nicotine, from *Nicotiana tabacum*) are able to "read" the sequences of symbols labelled as muscarinic and nicotinic receptors. We may assume that there are differences between nicotinic receptors located in the ganglia of the autonomous nervous system and the nicotinic receptors located in the motor end plates. The former are more specifically blocked by substances (drugs) that are called ganglioplegics (hexamethonium, tetraethyl ammonium, etc.), though nicotinic receptors located in the motor end plate are more specifically blocked by curares (natural and synthetic). Interpreting linguistically the action of the antagonists atropine, hexamethonium and curares, we may say that they are used strategically to reveal the language used by the organism to orient the molecule of the same mediator into three distinct morphological and physiological structures, namely the terminals of the parasympathetic system, the ganglia of the autonomous nervous system and the motor end plate. See page 000.

The situation is similar to that of the teacher writing on the blackboard three phrases with different meanings with the same homophonous symbols, and who indicates the desired meaning by simply erasing successively each one of the phrases or morphemes on the blackboard; so the Portuguese word *como* may signify the first person of the indicative of the verb *comer* (*to eat*), the Italian Lake *Como*, or the adverbial conjunction *como* (how *much*), and each meaning depends upon the context of the phrase:

"Como eu Como, no lago de Como."

How much I eat in the Lake of Como.

What can be explained to the student by the method of the successive eliminations: Como (*how much*), eu como (eat), in the Lake of Como, and so forth.

The problem for the organism is not as simple as that presented to the student of first degrees. When the organism endeavors to secrete gastric juices, to reduce the heartbeats, increase the movements of the digestive tract, it sends a message (mediator) that must decipher the code contained

---

*This double action of ACH, being a *nicotinic* substance, is not well understood by physicians and health agencies, who vehemently oppose the moderate use of cigarettes and cigars that might constitute a physiological stimulus for people deficient in the endogenous physiological function of acetylcholine.

in the muscarinic receptors; though when it (the organism) seeks to contract the striated muscles of a leg or of the neck, it sends the same messenger (mediator) that must "read" the code printed in the corresponding nicotinic receptors existing in the motor endplate; or, when it endeavors to stimulate all of the autonomous nervous system, it still is the same messenger, acetylcholine, which is sent to the right place, at the appropriate moment. As all that is done at the right place and the apropriate time, it depends upon the general integration by the central nervous system which uses the entelechia of the ancient philosophers, or the *wisdom of the body* of modern physiologists (W.B. Cannon). But in the ultimate analysis, all that should be done is to decipher the linguistic code that was printed in the synaptic membranes, to recognize the different forms (conformations) that can assume the molecule of a simple but versatile substance, such as acetylcholine.

Similar things happen in the case of nor-adrenaline and adrenaline (catecholamines) as mediators at the terminals of the sympathetic nervous system. There are functions of catecholamines that depend upon a "certain interpretation" of the reading of the code printed in the effectors of the sympathetic nervous system (acceleration of the heart, rise of arterial blood pressure, stoppage of the gastro-intestinal movements, dilatation of bronchioli, and so forth). Nor-adrenaline, which is the main mediator of the sympathetic nervous system, besides being able to recognize what is called an adrenergic receptor, in its general sense, must recognize sub-codes denoted as α_1, β_1, β_2... receptors.

This classification is based on the nature of the blocking agents (sympatholytic agents) of catecholamines, such as nor-adrenaline and adrenaline, in some of their effects and not in others.

When those effects refer to their action upon the smooth muscle of arterial vessels, producing a raise of arterial blood pressure, it is called an α-effect; when the observed effect refers to the acceleration of heartbeats, the effect is a β_1-effect, as first postulated by the American pharmacologist Ahlquist (1962); when the observed effect is the relaxation of bronchioli, the effect is a β_2-effect. However, when their effects upon the arterial blood pressure are blocked by an α-sympatholytic such as ergotamine or dibenamine, as first shown by Dale (1910) and Nickerson (1967), a vasodilator effect is revealed due especially to a β_2-effect produced by adrenaline.

Therefore, catecholamines must recognize (read) in the walls of the vessels two types of receptors, β_1 and α_1, the former responsible for a vaso-dilating, and the latter by a vaso-constricting action. The effects upon the heart, dependent upon the presence of β_1 receptors, are distinct from those existing in mammal bronchioli, upon which adrenaline produces a relaxing effect (β_2), and so forth. Such a variety of effects produced by the same mediator, or similar ones (adrenaline and nor-adrenaline) sent by the organism at the appropriate time to the appropriate locale of action, depends obviously on the capacity of such chemical agents to decipher in a fraction of a second, the sequence of symbols printed in the membranes of the effector organs. The clarification of such structures has been done, as in the case of acetylcholine, by the use of inhibitors (antagonists), that have to recognize by themselves such structures, the so-called α, β_1 and β_2 sympatholytic agents. However, also here, it is the *wisdom of the body (entelechia)* the superpower that decides the time and locale for the release of such physiological messengers, whether agonists or chemical mediators of the adrenomimetic type. (*See* Note III.)

Similar concepts could be emitted towards histamine, with its H_1 and H_2 receptors first postulated by H.O. Schild (1966), with their specific antagonists the classical anti-histaminics (anti-H_1), and the newly found anti-H_2 antagonists; or serotonin, the diversity of physiological actions observed would deserve a similar analysis of different receptors; or bradykinin which must activate at least four different kinds of receptors (Rocha e Silva, 1970); or prostaglandins, which act upon so many diversified structures; and all of those hormones and vitamins that must read, individually different morphemes of the complicated language of their *loci* of action in the living body, and so forth. Special books have been written about such linguistic problems, using of course the human language of the specialized scientist. See M. Rocha e Silva, *Fundamentos de Farmacologia*. Edart, Livr. Edit, (São Paulo), 1973-1979. Translated into Italian and Spanish (1978). See also A.B.

As mentioned previously, such considerations can be amplified to embody the mechanisms of action of the "classical hormones" (steroids, thyroxin, insulin, gonadotrophic and growth hormones, and so forth), the locus of action of which must be predetermined by means of a semantic code that must be recognized by chemical mediators to produce specific functions that constitute the main objective in the wide field of endocrinology.

The field of studies of such endogenous active substances that play such a fundamental role in normal and pathological physiology, was called by Henry H. Dale, celebrated English physiologist and pharmacologist, around the years 1920-1930, *autopharmacology* and became nowadays one of the most important chapters of pharmacology and of the utmost importance to physiology. In reality, all functions of the organism depending upon mediation by acetylcholine (cholinergic mediation), or by catecholamines (adrenergic mediation), or by serotonin (tryptaminergic), or by histamine (histaminergic), or by formation of polypeptides (bradykinin, angiotensin, substance-P) will fall under the title of *autopharmacology*. It is difficult to think of any function of the organism that would lie completely outside the domain of action of such mediators. Furthermore, if we include endocrinology into the ambitus of endogenous products with their specific receptors, as those for insulin, steroids, polypeptides (studied with techniques similar to those used for the simplest amines), practically all physiology (normal and pathological), psychology, therapeutics, become dependent on autopharmacological phenomena.

But, it is not this aspect, in a certain way "involving" of pharmacology, that raises so much the protest of physiologists, biochemists and psychologists, which we wish to stress. It is what exists of genuinely linguistics in deciphering the codes printed in the membranes of the living substance, and that are properly called *receptors*. If to that we add what has been said about the "enzymatic language" that was "invented" to be read by the specific substrate, we have altogether an extensive body of doctrine that covers practically all functioning of the living matter, from protozoa to man.

The other events or phenomena, electrical, ionic, physiocochemical, constitute at the ultimate analysis the mechanism by which the organism maintains the so-called *internal milieu* that we could interpret as supplying the material (ink, paper, glue, pen, photocopier, and so forth) used by the organism to print the semantic code to preserve the best conditions for its deciphering. But what seems most important, definitive, *sine qua non,* is the reading of the messages printed in the walls (membranes) of the living substance or in the macromolecules of enzymes that constitute part of the internal organization of the living matter.

Such "readings," code deciphering, semantic interpretations, is what can be denoted in a generical way as "exchange of information," or cybernetics of the biological process. The message to be deciphered is represented by the sequences of chemical groupings by the molecules of substrates, hormones, mediators and practically of all metabolites generated by the action of enzymes, and all that we could denote as the "structural grammar" of the biological language. Note that we still are in the phase of the so-called syntagmatic grammar, i.e., reduced to its phonetic and syntactics or, in other words, to "superficial structures" of the message transmitted or received by the living substance. What one could denote by "deep structure" (semantic) and transformational or creative grammar of the living substance, still sleeps in the prodigious history of natural evolution, initiated a few billions of years ago, in the lower-Cambrian, when a thin ribbon of DNA associated to a tenuous protein film, such as a primitive virus or "protobios" that already had the capacity of replication (or multiplication) inherent to the sequences of bases in the molecule of DNA, initiated the process.

Such a "big-bang" of the living substance some billions of years ago appears to us so complex that many biologists transfer the problem to other worlds or galaxies, assuming that some God-astronaut had come to our planet to seed life unities, after landing on some flat rock of the Cambrian. Others, more skeptical or inquisitive, try to explain the origins of life by means of simple reactions that nowadays can be reproduced in terrestrial laboratories. (*See* Note IV.)

Whatever could be or was such an origin of living matter, since its beginning the phenomenon consisted in deciphering a linguistic code, or codes, that now begin to be decoded or deciphered by human intelligence that already has conquered half a dozen or more Nobel Prizes. However, what has been achieved is no more than a simple stuttering of very simple concepts, when compared to the "Aristotelian Entelechia," the utterances of the God of the Bible, the *Elan vital* of Bergson, or the "wisdom of the body," of Cannon, that had the power to invent or decipher in the course of natural evolution, since the original plasmode (or protobios), that consisted of a DNA ribbon and a few chains of amino-acids, up to the complexity of the central nervous system of men in the nuclear era.

We have seen, in one of the previous chapters, that at a certain moment of evolution, living matter must have used a transformational grammar or language to read the code to transform the heavy annelid to acquire wings and conquer the aerial space; in a similar way that in more recent eras, a saurius (reptile) was transformed into a pterosaurius and then into a bird; or still a primitive lemurian to be transformed into a homid able to manipulate instruments and to emit sounds organized into an articulate language, very distinct from that emitted by their predecessors, and also completely different from that printed in the living substance itself. This last one is more similar to our printed language, developed much later, not earlier than about six thousand years ago. Only then, man acquired the capacity, previously solely inherent to the living matter, to print their messages on walls, small pieces of wood or ceramics, in papyrus, or paper, in such a way that others could decipher them.

The most significant example to our knowledge is that of the Rosetta stone, deciphered by Champollion in the beginning of the nineteenth century. For many centuries, up to 1806, the hieroglyphics of ancient Egypt (possibly the first form of phonetic writing employed by man) remained undeciphered, as a lost or forgotten language, to modern man. A merely fortuitous event allowed Champollion to decipher the code, profiting of the circumstance that phrases in Greek characters and Egyptian hieroglyphics coexisted in the same stone. The suspicion that the inscription was a bilingual printing of some Pharaonic code allowed a total deciphering of the hieroglyphics used by the Egyptians beginning in 3,500 B.C.

It was then verified that the hieroglyphics, though using pictograms, that is, pictures of known things, already had a phonetic structure in which the pictures had lost their pictoric value, and, as in human language, acquired the phonetic value of a conventional phonetic code to compose words (morphemes) or phrases and sentences in which the shape of the signal may have nothing to do with the meaning of the word or of the sentence (or phrase). So, the hieroglyphs of the word Cleopatra:

though using pictures of animals and things, the pictorical meaning of the same, has nothing to do with the meaning of the proper name Cleopatra. Incidentally, one may compare this sequence of hieroglyphs corresponding to the word Cleopatra with the sequences to be used by the organism to recognize the chemical mediators, acetylcholine, nor-adrenaline and histamine, as represented on page 95.

Another primitive language to be deciphered by linguists of the twentieth century, was the language registered in another type of hieroglyphics, that used by the Hittites, nomadic people living in the plains or deserts of the Middle East, a few thousand years before the Christian era. With great surprise, after deciphering the phonetic signs printed in wood and ceramics, it was verified that they used an Indo-European language, therefore of an origin similar to Greek, Latin and our modern languages. Therefore the connotation of *Indo-Hitita* given to the precursor of languages indicated before as *Indo-European.*

What we want to stress here is not the philologic or linguistic aspect of such sensational discoveries, but the fact that at a certain moment of his evolution, man started to write on the walls, on small tablets, stones or papyrus (or paper) using codes and symbols invented by himself, having profound analogies with codes and symbols employed by living matter to transmit its messages inside the living body.

As a last example of this extraordinary language of the living substance, we are going to give briefly the one that is printed in the "genetic code" in which context could be contained all planning of living matter, in its transformation from "protobios" to man of the spatial era.

The sensation that must have been felt by modern geneticists to decipher the "genetic code" must have been similar to that felt by Champollion deciphering the code of Egyptian hieroglyphics, or of the linguists of the nineteenth and early twentieth century deciphering the texts of Sumerians, Hittites and Phoenicians of the pre-Christian era. The *invention* of living substance consisted in finding in the sequence of bases (adenine, thymine, guanine and cytosine), in the molecules of DNA, to which is added uridine present in RNA, the possibilities of transmitting a practically infinite number of messages to the elaboration of protein macromolecules. In reality, the ultimate aim was that of deciphering the secret language of the "genes" (according to the *slogan* "to each enzyme a gene"), that is, unities existing in every cell of the organism. They are the molecules of DNA, immobilized in the chromosomes, the origin of the information brought by the molecules of RNA-messenger to the fixation or introduction of amino-acid molecules. Such molecules are fixed in the so-called RNA-transporter, which receives the order or information to carry them (each amino-acid is carried out by its specific carrier), to the line of assembling located in the plasma ribosomes.

If one considers that all the life of the cell depends upon the interactions of enzymes with their substrates and that the latter must be generated by the metabolism of precursors originated from food, or by enzymatic actions upon materials pre-existing in the living cell, we may have a glimpse of that fabulous organization of the living substance, that must receive the raw material and operate the transformation into agents, that by themselves will interact with molecules that constitute the general structure of the cell, receiving information or *blueprints*, messages, orders or whatever may be transmitted in the language of the "genetic code," depending upon the order or sequence of the former bases of the molecule of DNA, fixed upon the chromosomes (or genes).

Futhermore, we must also consider that it is the nature or chemical structure of DNA that determines the replication of the cell, in a perfect simile, in the cell reproduction. It is also in the reproduction of the organism by fusion of gametes that plays a fundamental role in the replication of chromosomes, or still in the discoordinated growth of cells in benign or malign tumors. All that must be produced in a certain degree of mental confusion to the non-specialist, but constitute the eternal source of dazzling to the biologist interested in the riddle of the origin of life.(*See* Note V).

A detailed and authorized description of the Genetic Code can be found in Luria (1973) and others mentioned in the Portuguese version of the book (Rocha e Silva, 1976). See also Additional Bibliography at the end of the volume.

Note I

The analogy we are trying to establish between the language of living matter and human language, or rather, between the grammar of Cartesian linguists (Chomsky and his adepts) with their finite number of rules, generating an infinite (or indefinite) number of phrases or concepts, in analogy with the infinite variety of natural events depending upon a finite number of laws or rules that have been discovered by scientists, in physics, in chemistry, in biology, or in the so-called human sciences, anthropology, sociology and so forth. Such an analogy can be extended to explain the way followed by the student or beginner to acquire their capacity to formulate scientific concepts (as the child learning its mother tongue), and eventually develop its creative capacity for scientific investigation. The main argument to accept such an analogy is the fact already stressed that the living substance itself in the course of its evolution along thousands or millions of years, has set a program in the form of a language printed in its membranes or in active centers of enzymes, just as the Egyptians acquired the habit of writing on the walls of their temples, in stones and papyri, messages, plans, endeavors in the form of hieroglyphics. Those who manifest their astonishment to the possibility of the intellect (or human unconsciousness) to contain anything preformed to receive the formal or transformational structure of the grammar of their mother tongue, understand (or admit) with the utmost naturalness that the fish or the amphibian had delineated the structure of a lung to the conquest of the extra-aquatic environment, with the extreme subtlety of the organization of a reptile, a bird, or a mammal, using a transformational language printed in their living matter. The analogy here suggested might contribute to orienting the teacher of science when trying to transmit to his pupils the complicated mechanism of the evolution of the species, the transmission of information in the living body, in the genetic code, in the *Aufbau* of the elements, in quantum mechanis, or in the laws of universal gravitation. Such results that usually must be memorized by the unhappy candidate to a place in the university (in Brazil) should be rather the object of a rediscovery in which the pupil should use its capacity of intuition, rather than its capacity of memorization or retention of frozen knowledge. Many teachers already do it that way and are considered *good teachers*. The majority, however, are unable to proceed like that, by ignorance or laziness, or by the resistance of the pupils themselves, already "conditioned" by many generations of bad teaching. Those responsible for programs of entrance examinations (vestibulares) and of preparatory courses (*cursinhos*), must be aware that to throw knowledge into the head of a student, without developing their capacity of creativity, is the same, or as useless, as trying to fill a barrel without a bottom, or with small holes everywhere. All of us who receive students in a more advanced phase of their formation (at the university) know that nothing remains of that enormous mass of knowledge introduced or memorized in the so-called cursinhos that infest Brazil, as malignant growths (cancers) that are corroding creativity or preparation of the university student. The example given here refers to the incredible process of choosing among the candidates (125,000), the 10% (12,000) who have a guaranteed place at the university. It seems, however, that the same problem appears in many other countries of the world.

Note II

To learn a new science should be like the "change of skin" of reptiles, amphibians and crustaceans in the course of their biological evolution. With the traditions of our bookish teaching that is merely informative, the student who begins its university courses already carries its rigid or impenetrable carapace, acquired in the so-called training for the entrance examination (vestibular). The latter instead of filling up the candidate's brain with badly assimilated knowledge should, on the contrary, stimulate "changes of skin," in such a way that the pupil would enter the university with the succulence of the soft crab, or of the crocodile that eliminated their juvenile shells, ready to start a new phase of growth and expansion. But even inside schools and faculties, or at least in *some* schools and faculties, the first years only contribute to reinforce the thickness of the carapace, instead of opening up new perspectives to the kind of teaching that is to come. In anatomy, biochemistry and physiology the student only acquires some rigid knowledge about the structure and functioning of the living body, in rigid or cemented programs prepared in advance, and no longer admits that everything can be seen (or regarded) in a different way, with the appeal to its intuition, namely its capacity for reading (or deciphering) the language of the living creature. In many faculties or schools (not to say the majority), physiology and biochemistry are understood as a frozen body of doctrine, to which the student only has to memorize the constituents of living matter, as well as the rules or laws that serve to prove practical finalities of the practice, in medicine, in pharmacy, or in dentistry, to mention only the branch of biomedical sciences. Rather than memorizing frozen knowledge, the student should learn the "language" of its science, written in the original papers that come out from laboratories all over the world. Excellent books, as mentioned in the references, are available that might fit such a purpose, in teaching physics, chemistry, biology (biochemistry, physiology, genetics and so forth), which are seldom, if ever, used as textbooks for the undergraduate or graduate student. Such a language of science is usually learned (or memorized) in compendia, or textbooks, only in its syntagmatic aspects, forgetting other books and papers of revision and original research, where the student or the future professional, is able to find the "transformational" aspects of the grammar of his science. To give a familiar example: One of the greatest difficulties of the student of pharmacology, which constitutes the first integrative branch of biomedical sciences, is understanding the extraordinary adaptation of physiological and pathological events to correlate the most complex functions of the organism based on the knowledge of the interplay of chemical mediators, in limited number, with receptors adapted or destined to them. It was such a difficulty that led physiologists of other epochs (though not very distant) to assume *vital forces* which were peculiar to the living organism. Though not all phenomena have been clarified by modern science, the natural tendency is to assume that such forces, which are not peculiar to living matter, all depend upon chemical and physical attractions and repulsions in accordance with quantum mechanics, in such a way that Chapter IV might constitute the natural introduction to the grammar of living matter, in what it has of the more profound or fundamental. What has received the name of *molecular pharmacology* is the study of distributions of electronic charges, atomic and molecular orbitals, charge transfers, that constitute the ultimate basis of understanding the interactions of drugs and chemical mediators with the living matter whose functioning is affected by them; and in biochemistry, the interactions of enzymes and substrates. But, still it is the integration of all such elementary forces in the harmonious ensemble, that is the living body, that is born, grows, reproduces and dies, which constitutes the greatest problem of modern biology. For the same reason, it is such a sector that constitutes the *milieu* of culture for the ideas of mystics and philosophers, in the old fashion. When we say that there exists a mentalistic or teleological part in the integration of all elementary phenomena that constitutes the harmony of the living body, this does not mean that we are projecting the laws and concepts generated in the human mind (or intelligence) to the real world. The latter has the primordial function to decipher laws and establish relations, in accordance with a grammar interiorized by man in the course of natural evolution, but it still is the object, the phenomenon, that must be the arbiter, or the incontestable judge of the veracity or viability of such laws and concepts. The apprenticeship of any science consists of trying to exteriorize the fabulous capacity of the human brain to formulate a limited number of concepts departing from a finite number of laws and rules, just as the child, in its first five years of age, exteriorizes its capacity to formulate new phrases, departing from a limited number of rules contained in the grammar of his mother tongue. I do not believe that the analogy between language and science can be brought much further. But of all human creations (or inventions), the language is that which must submit itself to *constraints* (or limitations) imposed by the laws of grammar of the language of the environment (*locus*) in which the child or the adult develop their living. Artistic creation, for instance, which also offers analogies with science in its creative phase does not impose such limitations (or constraints). The reality in which the artist lives, constitutes only a stimulation to the production of the work of art, and the limitations imposed to his creative power, are only the universal consensus that tend to classify the artists as good or bad. But not infrequently, he that has been considered one day a *bad* artist, can be reclassified as a genius. Example: The impressionists, cubists, dadaists (futurists) and whatever name has been given to the artists that have rebelled against the canons imposed by classical art. This subject has been discussed elsewhere.*

Note III

It is a truism to say today that pure scientific investigation is the basis for all applications of science. No example is better suited to the argument than that given by pharmacology, as the basis of therapeutics, that is the clinical application of drugs developed by pharmacologists. It was the observation of the existence of different receptors to acetylcholine, catecholamines, histamine and so forth, that led

---

*M. Rocha e Silva, *Lógica da Invenção e outros ensaios*, Livr. São José, Rio de Janeiro, 1965. And the dialogue with Anisio Teixeira (1968). *Diálogo sobre a Lógica do Conhecimento*. Edart Edit., São Paulo. The same subject has been extensively treated in a more recent publication. M. Rocha e Silva, *O Mito Cartesiano e outros ensaios*, HUCITEC Editora, São Paulo, 1978.

to the development of specific antagonists: atropinics, ganglioplegics, synthetic curares, sympatholytic agents, anti-histaminics and so forth. One of the objectives of such researches was to discover antagonists to chemical mediators, of possible use in practical medicine. Such an activity that might appear exceessively pragmatic that contributed on the other side to a better knowledge of the mechanism of action of the chemical mediators themselves. They are the agents that help to decipher the "language" existing in the pharmacological receptors.

Concerning receptors for acetylcholine, one might draw an arborized (tree) diagram, in a certain way analogous to those of the transformational linguistic, in that acetylcholine appears as an original (nodal) concept (or phrase) to which are subordinated all those in a hierarchized sequence:

Cholinergic transmission

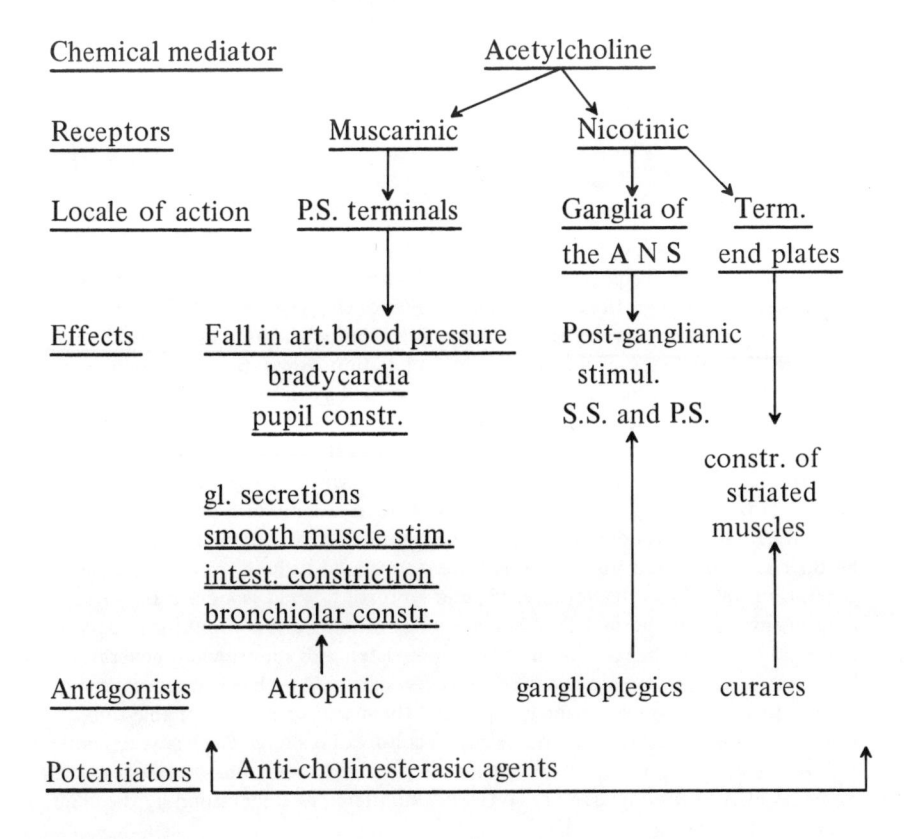

Chemical mediator	Acetylcholine	
Receptors	Muscarinic	Nicotinic
Locale of action	P.S. terminals	Ganglia of the A N S / Term. end plates
Effects	Fall in art. blood pressure / bradycardia / pupil constr. / gl. secretions / smooth muscle stim. / intest. constriction / bronchiolar constr.	Post-ganglianic stimul. S.S. and P.S. / constr. of striated muscles
Antagonists	Atropinic	ganglioplegics / curares
Potentiators	Anti-cholinesterasic agents	

Legend of Fig.

Syntagmatic analysis of cholinergic transmission in the autonomous nerve system (A N S = SS + PS) and in the terminal end plate of striated musculature.

The scheme presented in the accompanying figure of the cholinergic transmission has nothing of an esoteric or mystic nature. It is the usual scheme offered by a regular teacher to his pupils in what is called the elementary technique of teaching. Other phenomena of chemical mediation such as the biosynthesis of mediators, glandular secretion, inflammation, metabolization of mediators and of their antagonists, can be put in similar arborized or hierarchized schemes. By means of such schemes, homonyms or synonyms of scientific language can be identified by means of specific reagents that function as the "chalk and eraser" in any elementary class. In didactic books other syntagmatic schemes are represented, in which the analysis of scientific concepts are presented. Any good demonstration in class presents a succession of such schemes that lead finally to the apprenticeship of scientific language. The linguistic analogy can be found in schemes of "semantic selection" proposed by Katz and Fodor (1963) and Katz and Postal (1964) as cited in J. Greene, *Psycholinguistics: Chomsky and Psychology*. Penguin Books Inc., Middlesex, England, 1973.

Note IV

Once upon a time, spontaneous generation was an everyday accepted reality, and it was a well-known recipe proposed by Van Helmont (1577-1644) to produce rats, who believed it enough to lay down some pieces of cheese among old rags in some shaded corner of the house. It was also a common belief that the larvae of flies found in the carcasses of animals in decomposition originated by spontaneous generation. Such common notions, though contested by clarified spirits (minds), in eighteenth century, such as Spallanzani (1729-1799) who created the expression *omne vivum ex ovo,* have, however, persisted up to the middle of the last century even among highly ranked persons, and it was necessary to put up a strong fight by Pasteur (1822-1895) against his colleagues of the Academy of Science of Paris to prove the impossibility of spontaneous generation, even among such creatures that by their smallness and ubiquitousness were supposed to be born by this type of generation: infusories, bacteria, virus, and so forth.* What is significant in such long history is not the fact that humanity had needed a few thousands years of universal history to be convinced of the non-existence of spontaneous generation, but the opposite fact that in less than ten years, so rapidly, it has been convinced of the results of Pasteur's experiments, without the majority ever being able to see a flea egg, or observed through a microscope a culture of bacteria. Such is an example of how easily the achievements of science are received by the human mind, as the elements of spoken language are so easily assimilated or understood by the child who learns its mother tongue.

We could mention, still along this context of knowledge, in some way in opposition to individual experience, the non-existence of vital force, the movement of the earth around the sun, the law of conservation of energy, and in the

---

*See A. Delaunay, *Présence de Pasteur*, Livr. Fayard, Paris, 1973.

sociological or political fields, the recognition of the superiority of liberty over any form of tyranny. As far as spontaneous generation is concerned, everything would make one believe in its existence, starting with the appearance of life in the surface of our planet, assuming that some "cosmic wind" (radiation pressure) might have brought the germ of plants and primitive animals that came to continue their biological evolution in the billions of years that followed the upper-Cambrian. Nobody would be able to deny such a possibility, but, nowadays there is a tendency to explain or reproduce under laboratory conditions, the "protobios" that have been produced by the blind play of the hazard and quanta of energy of a highly photochemical power (that is, with the capacity of activating eventual molecules existing in the sea and ponds of the upper-Cambrian), originating molecules of amino-acids and organic bases to generate the first "double helix" of DNA, coupled to a ribbon of protein or polypeptides. It has been such a primitive small "worm" (minhoca) that initiated the course (arrancada) of the living substance in the following millions and billions of years. If there was in the upper-Cambrian some intelligent spirit who bent to the waters stagnated in the stones or crevasses that started to emerge from the sea, it could be able to see the inedited phenomenon, a swarm of small worms, swimming in increasing numbers, changing color to a green tonality, acquiring ameboid movements; and some millions of years later, the enormous variety of forms resulting from deciphering the codes of such transformational language obeying the laws of the grammar of living matter. And some billions of years later, the glorification of a small lemurian acquiring the shape of a human being, and the minimal conditions indispensable to dominate the whole creation, with its good and bad instincts.

Note V

The most extraordinary property of DNA is its capacity of replication, namely the capacity of a double helix, immobilized in the chromosomes, to suffer a longitudinal split (fissure) in two half-helixes, each one generating a similar structure by way of synthesis. This duplication occurs in chromosomes during cell division and constitutes a general phenomenon once the mass of DNA undergoes changes resulting from cell growth.

One of the most important results of the replication of chromosomes is the possibility of an equitable distribution of genes by the daughter cells resulting from cell division. In fact, if to each branch of the double helix corresponds the replicas resulting from the synthesis of a pair of DNA molecules, it is clear that to each daughter cell must correspond a replica of the ensemble of genes pre-existing in the chromosome of the mother cell. *Must correspond* is not the exact expression to be used, because the distribution is made entirely by chance, which makes the descendency (offsprings) diversified enough, though retaining the fundamental characteristics of father and mother, in the sexual reproduction. One can see here that the "endeavor" or "plan desire" of the living matter is that of assuring, as much as possible, an equitable distribution (in the meaning of a lottery) of their linguistic elements printed in the sequence of bases of DNA, what constitutes the essence of the *gene* where the "genetic code" is kept immobilized. When one deals with sexual reproduction in which the two gametes, father and mother, contribute with a quota-part to the formation of the fecundated egg, all that is mechanical, physico-chemical, molecular or quantical. The simple laws of hazard assure such an equitable distribution of the hereditary patrimony. If such was not the case, if there was no such equitable replication of the double helix constituents (genes), the world would be inhabited by repulsive monsters, many of them inviable, as if all animals and plants had received an overdose of antibiotics that would lead to the erroneous or tendentious reading of the genetic code, as in the thalidomide children, or bacterias that suffered the influence of streptomycin, and so forth. It is quite possible that such inviable monsters might have occurred among the "little worms" (*minhoquinhas*) of the upper-Cambrian, before the invention by nature of the double helix replication. The advantage of such speculations is that nobody can dispute them.

REFERENCES AND ADDITIONAL BIBLIOGRAPHY (AB)

REFERENCES AND ADDITIONAL BIBLIOGRAPHY (AB)

Ahlquist, R.P. 1962. The adrenotropic receptor-detector. *Arch. Intern. Pharmacodyn.* 139, 38.
Ahlquist, R.P. 1967. Development of the concept of *alpha* and *beta* adrenotropic receptors. *Ann. N. York Acad. Sci.* 139, 549
Ariëns, E. J., Edit. 1964. *Molecular Pharmacol. The mode of action of biologically active compounds,* Academic Press, N. York and London.
Ariëns, E.J. and Simonis, A.M. 1967. Cholinergic and anticholinergic drugs do they act on common receptors? *Ann. N. York Acad. Sci.* 144, 842.
Beadle, G. and Beadle, M. 1967. *The language of life.* A Doubleday Anchor Book, N. York.
Bergson, H. 1907. *L'évolution créatrice.*
Birks, J.B. 1963. *Rutherford at Manchester.* W.A. Benjamin Inc., N. York.
Bohr, N. 1958. *Reminiscences of the founder of nuclear science and of some developments based on his book.* Lecture given at the Physical Society of London. Nov. 28, 1958. Published in Birks (1963).
Bohr, N. 1961. *Physique atomique et connaissance.* Ed. Gouthier.
Cannon, W.B. 1945. *The way of an investigator. Autobiographical Assay.*
Castañeda, C. 1974. *The Journey to Ixtlan. The Lessons of Don Juan.* Penguin Books.
Castañeda, C. 1974. *The teachings of Don Juan. A Yaqui way to Knowledge.* Penguin Books.
Castañeda, C. 1976. *Separate Reality.* Pocket Books, N. York.
Chomsky, N. 1959. *Review of Skinner's verbal behaviour.* In: N. Chomsky (1959). *Language* 35, 28-58.
Chomsky, N. 1965. *Aspects of the Theory of Syntax.* MIT Press, Cambridge, Mass.
Chomsky, N. 1966. *Cartesian Linguistics.* Harper and Row, N.Y.
Chomsky, N. 1968. *Language and Mind.* Harcourt and Brace, N. York.
Chomsky, N. 1969. *La linguistique Cartésienne.* Le Seuil. Paris.

Chomsky, N. 1969. *L'Amérique et ses nouveaux Mandarins.* Cit. in Lyons, J. (1971). *Chomsky.* Edit Seghers, Paris.
Chomsky, N. 1971. *Linguagem e pensamento.* Edit. Vozes, Petropolis.
Chomsky, N. 1972. *Problems of knowledge and freedom.* Fontana, Collins, London.
Chomsky, N. 1974. In: New York Times. In Chomsky, N. (1974). *For Reasons of State.*
Chomsky, N. 1974. *For reasons of State.* Loc. cit. Chapter I.
Chomsky, N. 1975. *Reflections on Language.* Random House, N. York.
Chomsky, N. 1977. *Réflexions sur le langage.* Libr. F. Maspero, Paris.
Commoner, B. 1972. *The closing circle: nature, man and technology.* Bantam Edit. N. York.
Dale, H.H. 1910. Cit. in Dale (1933).
Dale, H.H. 1914. The action of certain esters and ethers of choline and their relation to muscarine. *J. Pharmacol. Exptl. Therap.* 6, 147.
Dale, H.H. 1933. Progress in Auto-pharmacology. *Bull. Johns Hopkins Hosp.* 53, 297.
Darwin, C. 1859. *The origin of the species.* See F. Darwin (1892).
Darwin, F., Ed. 1892. *The autobiography of Charles Darwin and Selected Letters,* Reprinted (1958), Dover Publ. Inc., N. York.
Engels, F. 1934. *Dialectics of nature.* Progress Publish, Moscow.
Fodor, J.A. and Katz, J.J. 1964. *The structure of language.* In: *Readings in the Philosophy of Language.* Edit by Englewood Cliffs, Prentice Hall, N. Jersey.
Frankel, C. 1973. The nature and source of Irrationalism. *Science* 180, 927.
Freud, S. 1935. *An autobiographical study.* Standard Ed., London.
Freud, S. 1972. *The Interpretation of dreams.* Avon Books, N. York.
Fromm, E. 1951. *The forgotten language.* Grove Press Inc. N.Y. (1957).
Fromm, E. 1971. *The crisis of Psychoanalysis.* Fawcett Premier Books, N. York.
Garaudy, R. 1966. *Pour connaître la pensée de Hegel.* Bordas, France.
Greene, J. 1973. *Psycholinguistics: Chomsky and Psychology.* Penguin Books, Inc., Middlesex, England.
Hall, J.L. 1978. Stabilized lasers and precision measurements. *Science* 202, 149.
Hegel, G.W.F. 1833. *Leçons sur l'histoire de la philosophie. La philosophie grecque.* (3 vols.). J. Vrin, Paris (1971).
Hegel, G.W. 1938. *Morceaux choisis,* Gallimard Edit. Paris.
Helmer, R. 1975. La lucha contra la contaminación del agua. *Cronica de la OMS* 29, 465.
Hoffman, B. 1974. An Einstein Paradox. *Trans. N.Y. Acad. of Sciences* 36, 721.
Huxley, A. 1954. *The doors of Perception.* See also: *Collected essays* (1966). A Bantam Book. N. York.
Huxley, A. 1957. The story of Tension. *Ann. N.Y. Acad. Sci.,* 63, 675.
Hyppolite, J. 1968. *Introduction à la Philosophie de l'histoire de Hegel.* M. Rivière et Cie, Paris.
Jung, C.G. 1964. *The man and his symbols.* Dell Publish. Co., N. York.
Jung, C.G. 1967. *Types psychologiques.* 3rd Edit. Buchet-Chastel.
Jung, C.G. 1970. *Psychologie et Alchémie.* Edit. Buchet-Chastel.

Jung, C.G. 1970. *Analytical Psychology: its Theory and Practice.* Vintage Books, N. York.
Jung, C.G. and Kerényi, C. 1971. *Essays on a science of Mythology.* Bollingen Series, Princeton Univ. Press, Princeton, N. Jersey.
Jung, C.G. 1973. *Approaching the unconscious.* In: *Man and his symbols.* A Laurel Edition. Dell Publish Co. Inc., N. York.
Karplus, M. and Porter, R.N. 1971. *Atoms and Molecules.* W.A. Benjamin, Inc., Menlo Park, California.
Katz, J.J. and Fodor, J.A. 1963. The structure of a semantic Theory. *Language* 39, 170. Reprinted in Fodor and Katz (1964).
Klemm, W.R. Edit. 1977. *Discovery Process in Modern Biology. People and Processes in Biological Discovery*, R.E. Krieger Publ. Co., Huntington, N. York.
Kretschemer, E. 1952. *Physique and Character.* Harcourt, Brace.
Kuhn, T.S. 1957. *The Copernician revolution.* Harvard, University Press, Cambridge, Mass.
Kuhn, T.S. 1971. *The Structure of Scientific Revolutions.* The University of Chicago Press.
Langacker, R.W. 1972. *A linguagem e sua estrutura.* Edit. Vozes, Petrópolis.
Leach, E. 1974. *Lévi-Strauss,* W. Collins and Cie Ltd., London.
Leakey, R. and Lewin, R. 1977. *Origins.* E. P. Dutton, N. York.
Leech, G. 1975. *Semantics.* Penguin Books, Middlesex, England.
Lefebvre, H. 1938. *Morceaux choisis de Hegel.* Gallimard, Paris.
Lefebvre, H. 1966. *Pour connaître la Pensée de Karl Marx.* Bordas, France.
Lenneberg, E.H. 1967. *Biological Foundations of Language.* John Wiley Edit., N. York.
Lévi-Strauss, C. 1970. *The savage mind. O pensamento selvagem.* (Trad. Edit. Nacional (1970), S. Paulo.
Luria, S.E. 1973. *Life. The unfinished experiment.* C. Scribner's Sons, N. York.
Lyons, J. 1971. *Chomsky.* Ed. Seghers, Paris.
Marx, K. et Engels, F. 1846. *L'idéologie Allemande.* Reedit Editions Sociales. Paris (1972).
Medvedev, Z.A. 1971. *The rise and fall of T.D. Lysenko.* Anchor Book Ed.
Michelson, A.A. and Morley, E.W. 1887. *Am. J. Science* 34, 333 In: *Brit. Encyclo.*
Murray, H.A. Edit. 1969. *Myth and Mythmaking.* Beacon Press, edition.
Nickerson, M. 1967. New developments in adrenergic blocking agents. *Ann. New York. Academ. Sci. 139*, 571.
Papaioannou, K. 1962. *Hegel.* Edit. Seghers, Paris.
Poincaré, H. 1902. *La Science et l'hypothèse.* Flammarion, Edit. Paris.
Regoli, D. and Barabé, J. 1980. Pharmacology of Bradykinin and related Kinins. *Pharmacological Reviews*, 32: 1-46.
Richards, J.P.G. and Williams, R.P. 1972. *Waves.* Penguin Books, London.
Rocha e Silva, M. 1965. *A lógica da invenção e outros ensaios.* Livr. São José, Rio de Janeiro.
Rocha e Silva, M. 1969. A Thermodynamic approach to problems of drugs antagonism. I. The "Charnière Theory." *Eur. J. of Pharmacol.* 6, 294.

Rocha e Silva, M. 1969. Concerning the histamine receptor (H_1). *J. Pharm. Pharmacol.* 21, 778.
Rocha e Silva, M. 1970. *Kinin Hormones.* Charles C. Thomas, Publisher, Springfield, Illinois.
Rocha e Silva, M. 1971. Semelhanças e dissemelhanças entre a criação na ciência e na arte. *Ciência e Cultura* 23, 2-7.
Rocha e Silva, M. 1972. *A evolução do pensamento científico.* Hucitec, Edit. São Paulo.
Rocha e Silva, M. 1973-79. *Fundamentos da Farmacologia.* Edart, Livraria Edit. São Paulo.
Rocha e Silva, M. 1974. Present Trends in Kinin Research. *Life Science* 15, 7.
Rocha e Silva, M. 1976. *Ciência Pura e Ciência Aplicada.* Hucitec, Edit. São Paulo.
Rocha e Silva, M. 1976. *Ciência, Technologia e Educação como base do desenvolvimento.* Reunião UDUAL, Mexico, Nov. 1976.
Rocha e Silva, M. 1978. *O Mito cartesiano e outros ensaios.* Hucitec Edit., São Paulo.
Rocha e Silva, M. 1978. A possible geometric interpretation of the electron jump in the hydrogenoid atom. *Spec. Sci. and Technol.* 1, 173-174.
Rocha e Silva, M. 1979. An example of a relativistic "red-shift" phenomenon in the hydrogen line spectrum. *Spec. Sci. and Technol.* 2, 555-563.
Rocha e Silva, M. 1979. A deterministic explanation of the electron jump in the hydrogenoid atom. *Ciência e Cultura* 31, 1136-1139.
Rocha e Silva, M. 1980. A relativistic equivalence principle for the electrostatic fields of charged particles. *Spec. Sci. and Technol. 3*, 177-184.
Rocha e Silva, M., Beraldo, W.T. and Rosenfeld, G. 1949. Bradykinin a new hypotensive and smooth muscle stimulating factor released from plasma globulin by snake venoms and by trypsin. *Amer. J. Physiol. 156*, 261.
Rutherford, E. 1909. See Birks (1963).
Schild, H.O. 1947. The use of drug antagonists for the identification and classification of drugs. *Brit. J. Pharmacology* 2, 251.
Schild, H.O. 1949. pA_2 and competitive drug antagonism. *Brit. J. Pharmacology* 4, 277.
Schilpp, P.A., Edit. 1957. *Albert Einstein.* Harper & Brothers. Publish. N. York.
Smellie, R.M.S. 1969. *A matter of Life—DNA.* Oliver & Boyd. Edinburgh.
Skinner, B.F. 1948. *Walden two.* The MacMillan Co., N. York.
Skinner, B.F. 1957. *Verbal Behaviour.* Appleton Century Crofts, N. York.
Skinner, B.F. 1972. *Beyond freedom and dignity.* A Bantam Vintage Book, N. York.
Teixeira, A. e Rocha e Silva, M. 1968. *Diálogo sobre a lógica do conhecimento.* Edart Livr. Edit., São Paulo.
Watson, J.D. 1968. *The Double Helix.* Weinfield and Nicholson, London.
Weber, M. In: Aron, R. 1967. *Les étapes de la Pensée sociologique.* Gallimard, Edit., Paris.
Weisskopf, V.P. 1975. Atoms, mountains and stars. A study of qualitative Physics. *Science* 187, 605.
Wittgenstein, L. 1921. *Tractatus logico-philosophicus.* Edit by J.A. Giannotti, Cia. Edit. Nacional. São Paulo (1963).
Wittgenstein, L. 1965. *Le cahier bleu et le cahier brun.* Gallimard, Paris.

INDEX OF NAMES

Index of Names

Adler, 25, 26
Ahlquist, R.P. 98
Ariëns, E.J. See AB
Aristotle, 10, 31
Aron, R., 55
Barrett, J. 80
Bacon, F. 34
Balmer, Series of 75, 78
Beadle, G. & Beadle, M. 40
Benchley, P. 7, 32
Bergson, H. 10, 101
Berkeley, G. 31
Bernard, C. 10
Bohr, N. 23, 72, 73, 74, 75, 77, 78, 79, 80, 81, 83, 89
Boltzman, L. 22
Born, M. 10
Boticelli, S. 21
Brackett (series) 78
Cannon, W.B. 98, 101
Castañeda, C. 17, 24, 25
Castro, Fidel 48
Champollion 101, 102
Chomsky, N. 8, 9, 10, 20, 23, 50, 51, 52, 53, 54, 58, 61, 68, 85, 86, 87, 103, 108
Churchill, W. 48
Cleopatra 102
Commoner, B. 6, 64
Copernico, N. 34, 43
Coulomb's, law. 73
Cuvier, G. 32, 39, 43, 72
Dale, H.H. 98, 99
Darwin, C. 10, 33, 38, 39, 40, 43
Darwin, F. 39, 43
Davisson and Germer 81

De Broglie, L. 76, 78, 79, 80, 81, 88
De Gaulle, C. 48
Delaunay, A. 108
Democritus 78, 84
Descartes, R. 10, 32
De Vries, H. 40
Diderot, D. 85
Dirac, P.A.M. 76
Dolfuss, E. 40
Einstein, A. 9, 10, 21, 74, 76, 77, 79, 80, 81, 88, 89
El Greco 21
Engels, F. 32, 33, 34, 35, 38, 39, 40, 41, 42, 48
Ferenczi, 26
Fodor, J.A. 9, 108
Fra-Angelico, 21
Franco, F. 33
Frank and Hertz 76, 77
Frankel, C. 20
Freud, S. 20, 21, 22, 25, 26, 27, 53, 62
Fromm, E. 21, 26, 62
Galileu Galilei 10, 33, 34, 42, 43
Garaudy, R. 35
Gassendi, 42
Goethe, J.W. 51, 54
Greene, J. 8, 108
Hall, J.H. see AB
Halley (Comet of) 16
Hanna, A.R. 80, 83, 84
Hegel, G.W.F. 10, 32, 33, 35, 36, 37, 38, 40, 41, 42, 47, 48
Heisenberg, W. 10, 76, 83
Heráclito 36, 37, 39
Helmer, R. 6
Helmont van 108

INDEX OF NAMES

Hitler, A. 33, 48
Hobbes, T. 22, 31
Hoffman, B. 10
Ho-Chí-Min 48
Humboldt, K.W. von 8
Hume, D. 10
Huxley, A. 25
Hyppolite, J. See A.B.
Jung, C.G. 20, 21, 22, 23, 25, 26, 27, 52, 54, 61, 62
Kant, E. 10, 24, 27
Karplus and Porter 75
Katz, J.J. 8, 9, 108
Kepler, J. 17
Kerenyi, G. 60, 61
Klemm, W.R. 24
Kohoutek (comet) 15, 16
Kretschemer, E. 51
Kuhn, T.S. 34
Langacker, R.W. 58
Laplace 34
Leach, E. 55
Leakey, R. 38
Leech, G. 9, 38
Leuwenhök, A. van 32, 72
Lefebvre, H. 36, 47
Leibnitz, G.W. 10, 27
Leite Lopes, J. 11
Lenneberg, E.H. 58
Levy-Brühl, L. 26
Levy-Strauss, C. 55
Leonardo, 21
Linneus, C. von 32, 72
Locke, J. 10, 22, 31
Lorentz, H.A. 43
Luria, S.E. 103
Lyell, C. 39, 43
Lyons, J. 23, 39, 43
Lysienko, T. 40, 42
Lyman (series of) 78
Mach, E. 10
Malinowski, B. 60
Malpighi, M. 32
Mao-Tse Tung 43, 48
Marx, K. 32, 33, 35, 36, 37, 39, 41, 48
Maxwell, J.C. 9, 10
Medvedev, Z.A. 42
Mendel, G. 40
Mendeleev, D.I. 84
Mendeleev, Table 71, 72, 76
Michelson, A.A.
Michelson and Morley exp. See AB
Michelangelo 21

Morgan, T.H. 40
Moses' Tables 93
Napoleon Bonaparte 33, 35, 36, 37
Murray, H.A. See AB
Newton, I. 17, 32, 34, 43, 57
Nickerson, M. 98
Nietsche, F. 51
Ovid 61
Pashen, series 78
Pasteur, L. 10, 108
Pauli, W. 76
Planck, M. 10, 21, 72, 74, 76, 77, 79, 81, 89
Planck constant 76, 77, 78, 80
Plato 10, 31
Plutarch 61
Poincaré, H. 10, 55
Postal, P.M. 10
Pythagoras 10, 55
Rank, O. 26
Regoli, D. See AB
Reich, W. 26
Richards, J.P.G. and R.P. Williams 82
Rimski-Korsakov, N. 81
Ritz' combination principle 89
Robinson, I. 8
Rocha e Silva, M. 19, 23, 38, 48, 55, 62, 77, 89, 90, 93, 99, 103, 106
Roeder, K.D. 62
Roosevelt, F.D. 48
Rutherford, E. 73, 75, 77
Rydberg, J.R. 75, 84, 88, 89
Rydberg (constant) 74, 75, 78
Schawlow, A.L. 76
Schelling, F.W.J. 60
Schild, H.O. 99
Schiller, F. 51
Schilpp, P.A. 89
Seuren, P.A.M. 8
Smellie, R.M.S. See AB
Schrödinger, E. 10, 76, 77, 78, 79, 82, 84, 88
Shakespeare, W. 51
Skinner, B.F. 20, 23, 50, 52, 58, 62
Spallanzani, L. 108
Stalin, J. 41, 42, 48
Teixeira, A. 106
Wallace, A.R. 43
Thomson, G.P. 81
Watson, J.D. See AB
Weber, M. 55
Weisskopf, V.P. 87, 88
Weyl's geometry 89, 90
Wittgenstein, L. 67, 69